MATTER
WITH ELECTROMAGNETIC RESONANCE

BY TIM WATERMAN

All information contained in this publication is derived from the author's independent research. Neither the author nor the publisher make any warranties of any kind, expressed or implied, with regard to the material or documentation included in this book. Neither the author nor the publisher shall be liable in any event for incidental or consequential damages in connection with, or arising out of, the furnishing, performance, or use of this material.

Unless otherwise noted, all text and photographs contained in this book are the property of the author. This book is not sponsored, endorsed, or otherwise affiliated with any of the companies whose products are represented herein. The opinions, beliefs, and viewpoints which are presented in this publication are expressed solely by the author.

Copyright © 2015 by Tim Waterman

This work is protected by both United States and International copyright laws and is provided solely for informational purposes. All rights reserved. No part of this book may be reproduced or transmitted in any form or by any means, electronic or mechanical, including photocopying, recording, or by any information storage and retrieval system, without permission in writing from the publisher. The scanning, uploading, and distribution of this book or any part thereof via the Internet or World Wide Web or by any other means without the expressed permission of the publisher is illegal and punishable by law.

Acknowledgements

A special thanks to Gary Evans,
my mentor as a young Engineer.

I would also like to thank my wife Debbie
for tolerating the two years of highs and lows
that came with writing this book.

Many thanks to my daughter Kelly
for the painting on the book cover.

Matter With Electromagnetic Resonance

Preface

Before I retired from the aerospace industry, I spent thirty-four years as an electrical engineer designing antennas and other electromagnetic devices. All of my designs were related to the interaction of electromagnetic waves and matter. I have always been an avid reader of science articles and books. In particular, I have been fascinated by the fact that when matter is annihilated, a large amount of electromagnetic energy is released. In my mind, that could not be a coincidence. Those thoughts have led me to write this book.

There are a number of important unanswered questions in the world of Physics. Why do the known particles have the masses and energies that they do? What exactly is Charge? What is Dark Matter? What is Dark Energy? Why is there more matter than anti-matter? The list goes on. In many cases, current theories do not begin to provide answers to these questions. This book presents a new theory which combines a containment mechanism and electromagnetic waves to produce what is commonly referred to

Matter With Electromagnetic Resonance

as matter. Matter comprises self-sustaining resonant electromagnetic cavity waves. Visually, the theory is simple and easy to understand. Particles and waves can be pictured using this theory. Particles are not abstract constructs, they are real things. This theory provides measurable, predictable answers to the questions posed above and many more.

Matter With Electromagnetic Resonance

Contents

Preface..	7
Introduction..	11
Chapter One - What's the Matter with Electromagnetic Resonance?..	24
Chapter Two - Bounce Modes and the Neutrino.............	39
Chapter Three - What do Atoms Look Like?......................	58
Chapter Four - The Nature of Galaxies and The Unification of Gravity, Electrostatic, Weak, and Strong Forces..	70
Chapter Five - Black Holes...	85
Chapter Six - The Aether..	96
Glossary...	104
Endnotes..	106

Matter With Electromagnetic Resonance

Introduction

I am writing this book for two audiences. My goal for the larger audience is to describe the physical world with words and pictures so the reader can visualize what electrons, protons and even black holes actually look like. For those readers with advanced degrees in science and math, I have included mathematical formulations to provide a solid basis from which to delve deeper into this theory. However, if I did my job correctly, these equations are not required to understand and appreciate the concepts presented here. An example of this approach might be explaining the concept of lift on a wing. I have seen many diagrams of wings with arrows on top and bottom representing different wind speeds. Statements are included about faster moving air on the top of the wing having lower pressure than slower moving air on the bottom of the wing. Greater pressure on the bottom gives the wing lift. Not a single equation is needed for an understanding of the concept of lift. If you were about to design a wing, however, equations become vital.

Matter With Electromagnetic Resonance

That matter comprises self-sustaining electromagnetic resonant cavity waves is not a new hypothesis, but I believe this is the first physical description of what particles look like, the containment mechanism and the subsequent properties of the predicted particles. Electromagnetic resonant cavities are fixed-size structures which are capable of storing oscillating electromagnetic energy at frequencies related to the size of the structure. A vibrating guitar string is an example of a mechanical resonant structure. The string can store many vibrating resonances at different frequencies. A microwave oven is an example of an electromagnetic resonant cavity. Microwave frequency energy bounces around and is contained inside the oven cavity (eventually heating the food).

To get an idea of how electromagnetic resonance might work to make matter, I start by looking at the electrons and protons around us. Electrons and protons are known particles. Protons have mass and volume (although the value of the volume is subject to some speculation, there is some kind of value). Electrons have mass. A safe thing to say about the volume of an electron using current theory is that it is unknown. Today electrons are called "point like", but that is not the same as having zero volume. A zero volume particle with mass would have an infinite density. This seems unlikely to me. For this theory I am assuming that the electron has a volume. Because the definition of density is mass divided by volume I can say that these particles have a finite density. By measurement I can also say that they have equal and opposite charges "+/- Q". Charge "Q" is defined as the property of

Matter With Electromagnetic Resonance

matter which causes it to experience forces in the presence of other charged particles. I may not know exactly what mechanism generates the charges, but their values have been measured.

I know that an applied voltage "V" is capable of moving these dense charges around. Moving charges are called current "I". Because voltage is capable of moving density from one place to another or increasing density in one location at the expense of another, voltage is considered to be a pressure.

What is the connection to the creation of particles? Since voltage pressure in the world around me is capable of moving density, it suggests to me that charged dense particles themselves might be generated by voltage. Perhaps excess space density is compressed and maintained by voltage to create particles. For any particle to exist for an appreciable amount of time, the voltage pressure would have to always be present and be stable. I believe a lossless resonant cavity of some kind would be capable of containing the voltage pressure as well as the compressed excess space density. Remember, a resonant cavity is a structure which is capable of storing oscillating energy.

This book proposes that space is a material with elastic and inelastic properties and that voltage has the ability to compress space and increase its density. This book also proposes that the resonant cavity structure itself is created by voltage pressure. I will describe how spherically shaped electromagnetic resonant cavities are created and sustained by the voltages of the contained electromagnetic wave. This new theory will explain the electron, the proton, the neutron, and much more.

Matter With Electromagnetic Resonance

Some of the surprising possibilities, explanations and ramifications of this theory include:

- All stable particles are spherical bubbles with different wall thicknesses, peak densities, masses, energies and resonant frequencies
- The actual size and shape of electrons, protons and neutrons can be predicted
- Explains what the likely quarks are in a proton
- Predicts the existence of Dark Matter/Dark Energy, its size shape and other properties
- Explains why anti-matter is less stable than matter
- Explains why the speed of light is constant and has hit its maximum speed today
- Explains why there are Supernovas "Dark Neutrons"
- Explains what neutrinos are and their properties
- Provides a simple mechanism which explains how atoms are put together
- Allows a simple understanding of what gravity is
- Explains where galaxies come from
- Unifies the four known forces [gravity, nuclear strong, nuclear weak, electrostatic]
- Explains supermassive black holes

Starting with my basic approach, the energy containment

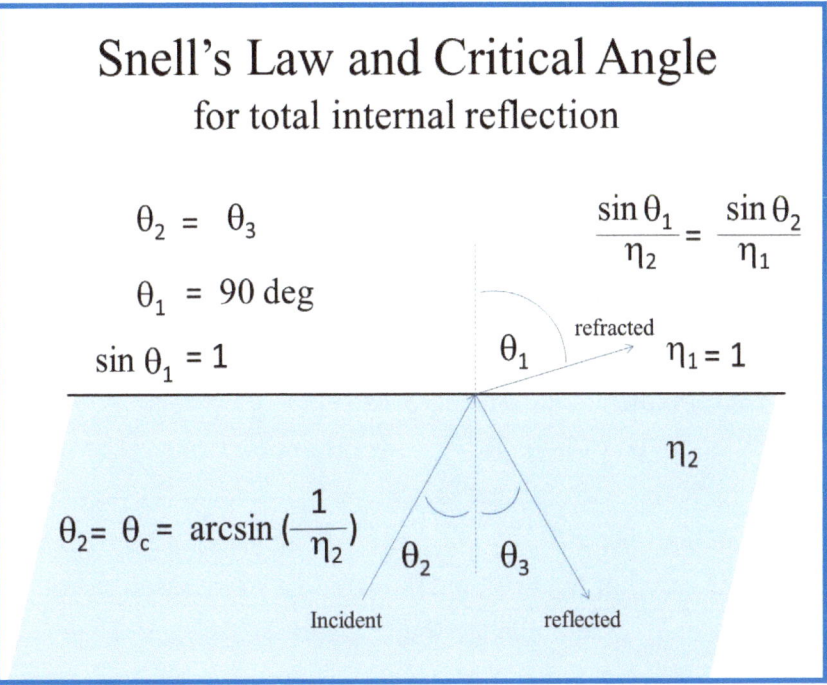

Figure A.1 Snell's Law as it pertains to the "total internal reflection" of electromagnetic waves at the boundary between two media with different indices of refraction (η_1 and η_2). The sine of the critical angle (θ_c) for total reflection is dependent upon the ratio of the two indices. The greater the index inside, the closer the critical angle is to zero. The critical angle can be zero (normal incidence) only if the index of refraction on the outside (η_1) is zero.

mechanism for a self-sustaining electromagnetic resonant cavity is Snell's Law as it pertains to the total internal reflection of an electromagnetic wave at the boundary between two media.

Referring to Figure A.1, for conventional materials, imagine a change in the index of refraction (η) at their common boundary. The index of refraction (η) is the measure of the speed of light in any medium compared to a reference medium. The standard reference medium used today is "free space" for which the

Matter With Electromagnetic Resonance

assigned reference index of refraction (η) is 1.0. When the ratios of the indices of refraction (η) are specific values then total internal reflection can occur at critical angles. [$\theta_{incident} = \theta_{critical}$ = arcsin (η_1/η_2)]. The premise for the self-sustaining electromagnetic cavities is that when the voltage of an electromagnetic wave is high enough, it can compress space (make it denser) and slow down the speed of light in that compressed space. (Examples of the speed of light slowing down in dense materials are all around us. Glass, plastic and water are examples of materials through which the speed of light slows down.) In other words, the voltage pressure of the wave squeezes space and increases the density of space in the direction of the pressure. Increased density slows down the speed of light passing through that compressed space.

How can voltage change the density of space? Voltage is a pressure. When the pressure wave (voltage) gets high enough it can push and squeeze together the medium upon which it is traveling. If the medium is compressed by a factor of 2 then (for the same frequency) the wavelength shrinks by a factor of 2. Since the speed of the wave (C) is equal to the product of the frequency times the wavelength, the speed is reduced by a factor of 2.

The density function for space and how it relates to the index of refraction (η), frequency, particle size and mass can be derived from the known laws of physics. (The derivation of the formula for visible matter is shown on pages 36-38). This new theory of voltage pressure changing the density of space, a space which has both elastic and inelastic properties, explains much of what is not understood about the nature of matter and the universe.

Matter With Electromagnetic Resonance

There is a lot of anecdotal evidence that particles are spherical in shape. So as not to reinvent the wheel, I have focused largely on spherical resonances. In Chapter Four, I will discuss the possibility of cubic resonances in the early universe.

The Self-Sustaining Electromagnetic Resonant Cavity Modes

The idea for possible stable spherical cavity mode shapes and containment mechanisms arose from a need to match some known quantities. For instance, protons and electrons have equal and opposite charges (+/- Q). Neutrons have no apparent charge. They all have spin and mass, although the masses of the proton and electron are quite different from one another. This is likely due to them having different resonant frequencies based on the energy relationships in Equation 1.1.

$$e = m \cdot c^2 = h \cdot f \qquad \text{(Eq. 1.1)}$$

e = *Energy (joules)*
m = *Mass (kg — kilograms)*
c = *Speed of light in free space (3×10^8 meters sec^{-1})*
h = *Planck's constant (6.626×10^{-34} $meter^2$ kg sec^{-1})*
f = *Frequency (Hz)*

The proton and neutron are close in mass which suggests that their frequencies are similar.

Matter With Electromagnetic Resonance

Figure A.2 shows the derivation of a possible spherical mode shape which can produce a spinning charge. A traveling electromagnetic wave in free space might be depicted similar to the neutral picture (a) in Figure A.2. The voltage oscillates between positive and negative as it travels through space. If I was able to ground one side of this waveform, then all of the voltage changes would either be positive or negative. This can be seen in the *positron* and *electron* (b and c) views of Figure A.2. The last

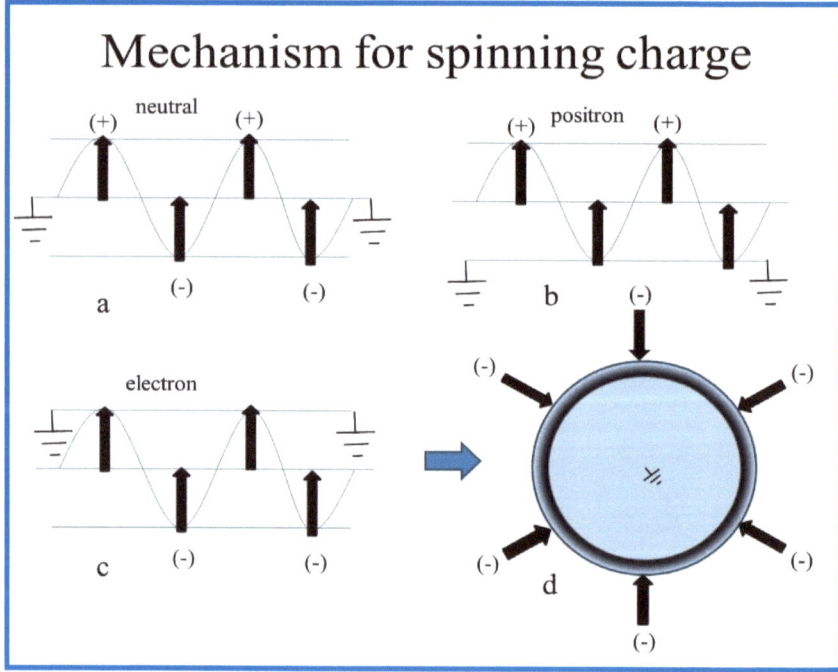

Fig A.2 A mechanism that can generate and sustain a fixed spinning charge is a resonant electromagnetic spherical cavity where the polarities on the opposite sides of the inside of the cavity cancel one another at the center forming a virtual ground. The virtual ground on the inside forces the outside of the resonant cavity to either be positive or negative at all times. a) Neutral is grounded in the middle of the waveform. b) Positron is grounded on the negative pressure side. c) Electron is grounded on the positive pressure side. d) The travelling electron waveform wrapped in a resonant spherical cavity with negative charge pointing outward.

Matter With Electromagnetic Resonance

view (d) shows how the *electron* travelling wave would look within the confines of a circle (or enclosed in a sphere).

This electric field shape makes its own virtual ground in the center of the circle caused by the cancelling voltages on opposite sides of the center. Depending on which way the voltage is pointing (the direction of pressure), one can see by inspection that a permanent (+) or (-) charge is presented to the outside world. This is the shape of the simplest resonant cavity. In the first chapter I will discuss why, for this simplest mode, the negatively charged particle (*electron*) is more stable than the positively charged particle (*positron*). This theory explains why anti-matter is not abundant today in the universe.

How to Use Snell's Law for Total Internal Reflection in Spheres and its Relation to the Density of Space

To simplify the visualization of the different stable spherical resonances and how they can sustain themselves, it is easier to look at a cross-section through the centerline of a resonant sphere. Figure A.3 shows many possible stable cases (and a few unstable). In the top two pictures in Figure A.3 (a and b), using Snell's law, the angle of incidence (critical angle) for the contained electromagnetic wave is equal to 90 degrees (grazing). This is the angle between any tangent line at any point around the sphere and its intersecting radial line. Total internal reflection is achieved with only a gradual increase in density as one moves toward the cavity center. I call these graded densities, where the

Matter With Electromagnetic Resonance

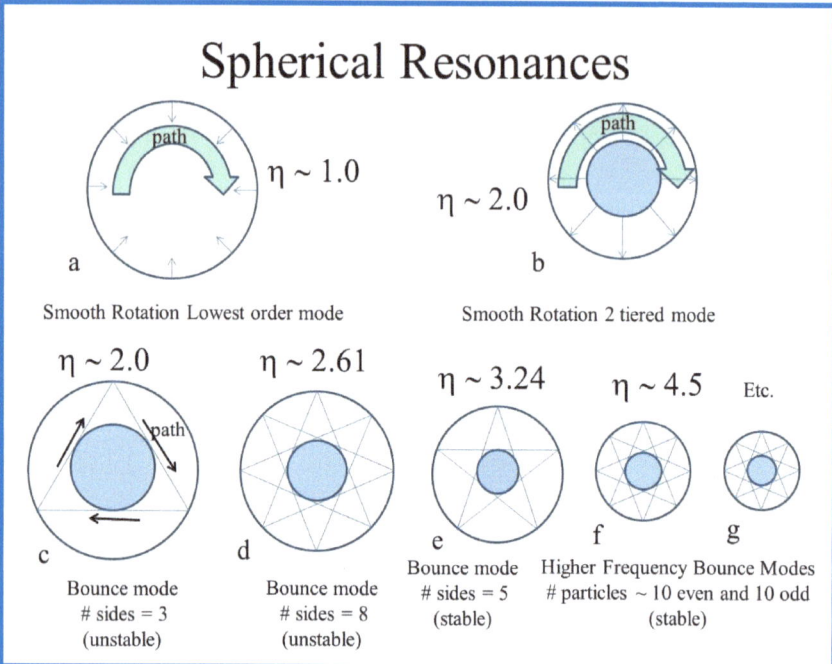

Fig A.3 Allowable spherical resonant mode shapes as predicted using Snell's Law for total internal reflection. With the electric field polarized radially, visible matter comprises the top two mode shapes (a and b). For these modes the electromagnetic wave spins smoothly around a central point or virtual ground. With the electric field polarized circumferentially Dark Matter (as well as Dark Energy) comprise the bottom row of modes (c,d,e,f,g). I have named these shapes Bounce Modes because of the way the wave bounces around the inside of the sphere. The value of the index of refraction (η) in each picture comes directly from Snell's Law and is used to calculate the resonant frequency and peak mass of each particle.

index of refraction (η) increases linearly toward the center, inversely proportional to the radius.

The upper left picture of Figure A.3 uses small arrows to indicate the direction of increasing voltage (pressure). If the density of space changes smoothly and increases linearly toward

Matter With Electromagnetic Resonance

the center of the circle then that same electromagnetic wave that is increasing the density can rotate perfectly in a circle and be contained. (It is the same for a sphere if you consider the other planes doing something similar). Each circumferential path, no matter what radius, has exactly the same electrical length. Each length is modified by the change in the speed of light due to the increasing density of space toward the center. In a way, the travelling wave thinks it is going in a straight line, but because of the increased density of space toward the center and the shrinking wavelength, it is actually travelling in a circle. This is the case for the electron, proton and neutron although the proton and neutron consist of a more complex two-tiered higher order mode as shown on figure A.3(b). The particles in the bottom row of Figure A.3 (c,d,e,f,g) are polarized differently than visible matter (a and b). I call these bounce modes because of the bouncing path that the electromagnetic wave takes through the particle. Each of these bounce modes has a different stability. The most stable particles are candidates for Dark Matter. The least stable are candidates for Dark Energy.

Voltage (pressure) induced density of space changes, alter the speed of light where the voltage is present in every example shown in Figure A.3. Another requirement for these spherical cavity modes to be stable is that the resonant sphere diameter must be between $(3/\pi)$ Lambda (wavelengths) and 1.0 Lambda in whatever index of refraction (η) is present. The lower bound puts the circumference at 3 wavelengths with a radius just under ½ Lambda. If the diameter is greater than 1.0 Lambda, the upper

Matter With Electromagnetic Resonance

bound, then higher order modes can exist and it is no longer the stable simple mode.

The following are samples of what will be discussed in the book.

Chapter One - In chapter I visible matter particles will be discussed. What do they look like? What are some of their properties?

Chapter Two - In chapter II Bounce Mode particles will be discussed. What do they look like? What are some of their properties? Are these particles Dark Matter? Is there any measured data to support their existence? Last, a description of the neutrino will be provided that comes from this theory. The neutrino's self-sustaining mechanism is different than the mechanism used by the spin or bounce modes, but still has to do with the compressed volume of space "Q" and its density.

Chapter Three - The assembly of atoms from the known visible particles, electrons, protons and neutrons is cast in a brand new light with the realization that these fundamental particles are self-sustaining electromagnetic resonant density bubbles with predictable sizes, masses, charge distributions, and other properties. The relationship presented in this book between allowable densities of space (indices of refraction) and resonant frequencies for stable particles enables simple calculations to be performed that accurately predict the sizes of all of the atoms in the periodic table. I will describe in simple, easy-to-visualize terms the actual arrangement of electrons, protons, and neutrons inside all atoms. This actual arrangement is quite different than

Matter With Electromagnetic Resonance

how atoms have been depicted up until now. This new description offers an easy-to-understand perspective that I believe has clear logical advantages over the existing theories of what atoms look like.

Chapter Four - The possible origin of galaxies is provided in this chapter. Unification of the four known forces of nature is possible using this new theory and will be presented.

Chapter Five - The idea that super-massive black holes are present at the center of most, if not all of the galaxies, is a natural outcome of this new theory. Physical sizes, masses and the different gravitational strengths of black holes are presented.

Chapter Six - A description of a material will be presented. The material's size and shape and other properties, suggest to me that it might be the Aether.

Chapter One
What's the Matter with Electromagnetic Resonance?

Visible Matter

All of the stable particles predicted by this theory are in fact bubbles of excess space density. Voltage is simply the trapped pressure inside of the self-sustaining three dimensional bubbles. Charge is one side or the other of this voltage pressure gradient as it squeezes a fixed amount of space (~1.0Q) into a denser material. Current is the movement of these excess density bubbles and indeed the movement of the excess density contained in the bubbles as the density wave propagates in a fixed medium (space or the Aether). In the descriptions of these particles, I use the terms voltage and pressure interchangeably. Electromagnetic

Matter With Electromagnetic Resonance

wave and density wave are also synonymous. The index of refraction (η) and the density of space are also used interchangeably. "Density" as it is commonly used in physics (mass ÷ volume) will be used when talking about the relationship between the peak index of refraction and the peak mass or resonant frequency of each stable particle.

Referring to Figure 1.1, which is a cross-section of a proton (with a peak index of refraction (η) = 2.0). For an electromagnetic plane wave to rotate around in this perfect circle, the speed C of the electromagnetic wave must decrease linearly and be proportional to the radius of the circle. For example; a circumferential path ½ ways in from the outside perimeter must have a density of space (index of refraction) that is twice that of the perimeter. The speed of the wave at that inner circumference will be ½ that of the outside. This smoothly increasing density of space allows the entire plane wave to make a complete rotation about a single point (the central virtual ground). This rotating wave eventually ends up back at its starting point to add in phase with itself. To be resonant, the total path circumference must be an integer multiple of a wavelength. For the visible particles these criteria are met when the outer diameter is (3/pi · Lambda) at the resonant frequency. The circumference is three wavelengths. It should be noted that since the wavelength shrinks linearly with the density, every circumference inside the resonant circle (or sphere) has a diameter of (3/pi · Lambda) in the index of refraction present at that radius.

Matter With Electromagnetic Resonance

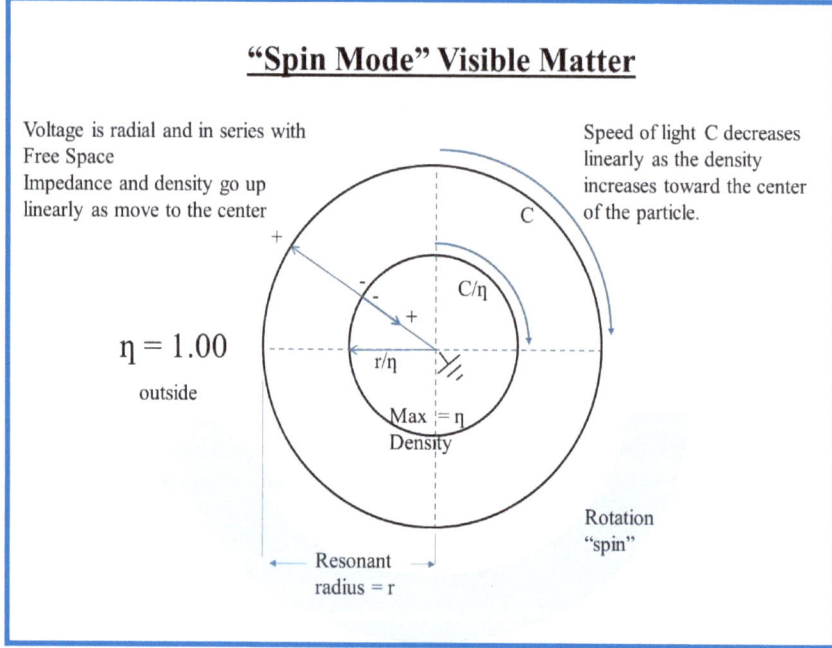

Fig 1.1 Spin Mode particle (proton), three wavelengths in circumference. The speed of light "C" decreases linearly as the density of space (η) increases linearly toward the center.

The Density Function for Spin Modes "Visible Matter"

Using the standard laws of physics for energy, mass and volume along with Ohm's law for the relationship between voltage and current (moving charge Q), I can calculate the relationship between the maximum density of space (peak index of refraction "η"), voltage, mass and frequency. The answers are shown in Equations 1.2 and 1.3. The peak index of refraction for the simplest mode electron ($η_0$) is so close to 1.0 that the electron

Matter With Electromagnetic Resonance

mass and frequency can be used as $Mass_0$ and $Freq_0$ to calculate the masses and frequencies of the other stable particles. Remember that the indices of refraction are all relative to one another.

> $Mass_1 = Q_1^2 \cdot \eta_1^{10} \cdot Mass_0$ where $Mass_0 = 9.1 \times 10^{-31}$ kg
> (electron mass) (Eq. 1.2)
>
> *And because of the relationship between mass-energy and resonant frequency-energy*
>
> $Freq_1 = Q_1^2 \cdot \eta_1^{10} \cdot Freq_0$ where $Freq_0 = 1.236 \times 10^{20}$ Hz
> (electron frequency) (Eq. 1.3)
>
> Where $Mass_1$ = Peak mass for new particle #1
> Q_1 = Peak amount of space being compressed in particle #1
> η_1 = Peak index of refraction of that compressed space
> $Freq_1$ = Resonant frequency for new particle #1

This relationship holds for frequencies near or below that of the proton (2.25×10^{23} Hz). Although the charge density can change in this frequency region, the actual net charge Q is constant. This is why the charge on an electron and the net charge on a proton are equal in magnitude but opposite in sign. The derivation of this function is shown on pages (36-38).

Figure 1.2 (*next page*) shows the calculated density function of space vs. frequency for visible matter. The approximate locations of the electron and proton are shown. These particles are stable at those frequencies.

Matter With Electromagnetic Resonance

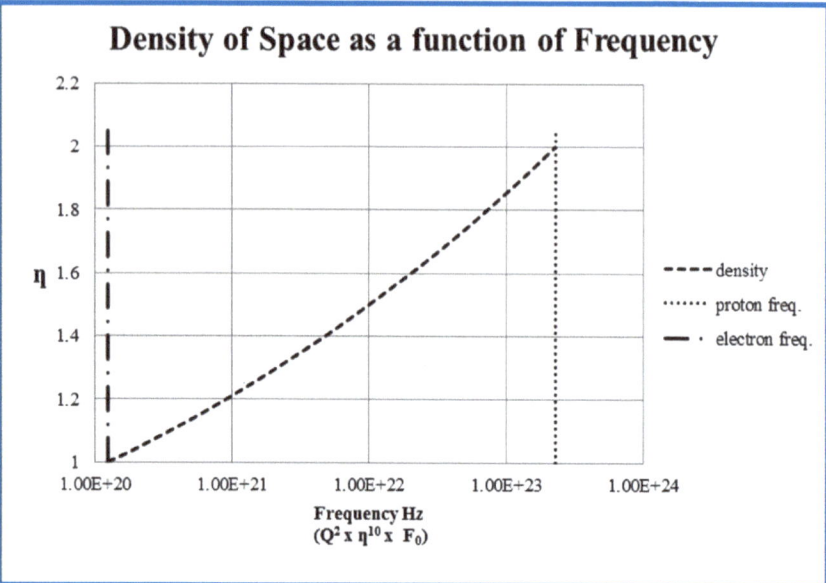

Fig 1.2 *The density of space (index of refraction) as a function of frequency. Using the electron as the fundamental frequency (f_0) where the index of refraction is ~ 1, at the proton frequency (2.25 x 10^{23} Hz) the peak index of refraction is equal to 2. For the purposes of this graph the peak amount of compressed space Q has been allowed to vary smoothly from 1.0 at the electron frequency to 1.333 at the proton frequency.*

The Electron

For a more in-depth look at stable visible matter, the electron appears to have the simplest mode shape, a single spherical cavity ~.955 Lambdas (wavelengths at its resonant frequency) in diameter. In the frequency domain (1.236 x 10^{20}Hz $=f_0$) the electron exists where the increase in the index of refraction caused by the wave's voltage exactly matches the change in radius across the wave-front as the wave rotates around the center

Matter With Electromagnetic Resonance

of its sphere. The electron's peak index of refraction is slightly greater than 1.0 (~ 1.0 + 1.0 x 10^{-10}). This is based on the size of the electron (radius ~ 1.16 x 10^{-12} meters) and the volume of charge contained (charge Q ~ 0.70 x 10^{-45} meters^3). Because the electron charge volume is the same as the net charge volume of a proton (whose resonant cavity is physically smaller). When this fixed volume is spread out to the radius of the lower frequency electron, it takes the shape of an incredibly thin walled spherical bubble. The bubble wall thickness is ~1.0 x 10^{-22} meters. Because the peak space density ($\eta_{electron}$) caused by the voltage of the electron is close to 1.0, for the purposes of this book the electron frequency will be used as the fundamental frequency (f_0) against which all other densities and frequencies will be calculated.

Figure 1.3 (*next page*) shows a top view of the electron. In this plane the electromagnetic wave is a traveling wave. This rotation is probably what provides the property called "spin" in particle theory. Because the voltage gradient points inward, the densities (and voltage) increase toward the center. In the other two planes of the sphere, which has a similar voltage profile, there is probably a standing wave, the sum of two waves traveling in opposite directions (although I believe they could be spinning in those planes as well having a net diagonal spin). Because positive pressure is directed toward the center, it is repelled by the opposite side of the sphere which is also positive (like charges repel, opposites attract). This repulsion is what forces the compressed space of the electron to be a thin shell structure. The electron bubble wall thickness is much thinner than a wavelength

Matter With Electromagnetic Resonance

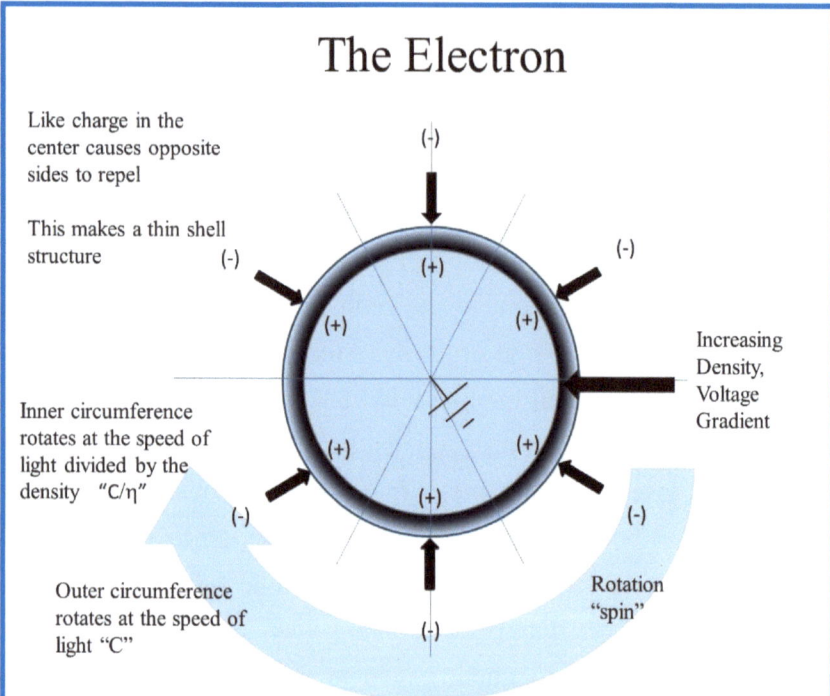

Fig 1.3 A slice through the spinning electromagnetic wave that comprises the electron. The voltage pressure increases toward the inside increasing the density of space ever so slightly toward the center allowing the wave to travel in a perfect circle.

at the electron frequency. Based on this thickness, the calculated peak index of refraction for the electron works out to be (~ 1.0 + 1.0 x 10^{-10}).

The Positron

The description of the electron above raises the question of the possibility of a positron (anti-matter), when the charge and increasing pressure (voltage) are positive on the outside. What

Matter With Electromagnetic Resonance

would happen if the voltage gradient points out? This would tend to make the density greater toward the outside, equivalent to a frequency that is just below the fundamental frequency (f_0). This particle is not stable. The voltage increase toward the outside forces the charge Q on the outside to be a little bit thicker than that frequency can support. (*See Figure 1.4 on next page*)

When a positron comes into contact with an electron they will annihilate each other, but because the shells are extremely thin (~5.67 x 10^{-11} Lambda) annihilation will take a finite amount of time which can be calculated.

If it takes 1.76 x 10^{10} cycles for the positron shell to spin off (1.0 / 5.67 x 10^{-11} Lambda [shell thickness]), at 1.236 x 10^{20} Hz it will take 1.427 x 10^{-10} seconds. This is very similar to the published positron annihilation times (1.0 x 10^{-10} to 4.0 x 10^{-10} seconds) in different materials.[1]

I can also estimate the lifetime of a positron without annihilation. I could use a similar technique as above, but figure out how long it would take for the "extra Q thickness" to spin off. The extra thickness that causes instability works out to be ~10^{-33} Lambda. At 1.236 x 10^{20} Hz it would take 256,000 years to spin off. Another estimate treats the extra thickness like a small series resistor. It would take 177,000,000 years to burn up all of the charge. In either case the times are significantly shorter than the life of the universe, which would explain why there are more electrons than positrons.

Matter With Electromagnetic Resonance

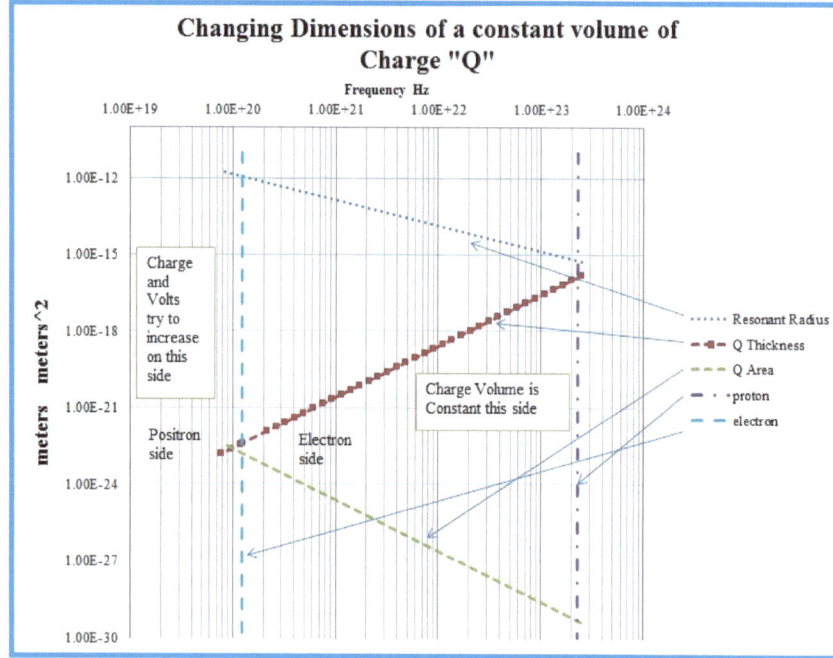

Fig 1.4 The changing dimensions for a fixed bubble volume of space (charge Q ~ 0.70 x 10^{-45} meters³). As the resonant radius increases, the thickness of the bubble wall decreases. At the electron frequency the spinning bubble is stable with a negative charge outside. Positive pressure pushing out as in a positron is invariably unstable. The positron mode will disintegrate over time depending on the boundary conditions outside of the positron. At this slightly larger radius the increasing area of Q goes up faster than voltage. This decreases the index of refraction (η< 1) on the outside just enough to add some short-term stability. The index of refraction on the outside is (1.0 - 1 x 10^{-10}). However, it is not as stable as the electron which has the voltage gradient pointing inward. The positron has a charge Q that is slightly greater than free space allows. This extra constraint is what I believe makes the positron disintegrate more easily than an electron. There is dissipation of the extra charge over time into free space because the particle is trying to rotate faster than the speed of light.

Matter With Electromagnetic Resonance

The Proton and Neutron

These denser particles must be more complex than the electron. The proton is a stable spin mode particle with a net charge of +1Q. Figure 1.5 (*next page*) shows the shape of this two-tiered structure. There is a sinusoidal varying electromagnetic wave spinning around the outer tier.

The inner tier comprises another spinning wave polarized in the opposite direction. The inner tier has the polarization and shape of the simple electron mode. The peak space density (η) for this stable particle is 2.0 and occurs at the boundary between the two tiers. Using the equation for visible matter (Eq. 1.3), the frequency at which this space density (2.0) occurs is $1.333^2 \times 2.0^{10} \times 1.236 \times 10^{20}$ Hz = 2.25×10^{23} Hz. For this positive outward pointing pressure to be stable (the direction of the voltage gradient looks like the unstable positron) something has to happen to the voltage, charge and density as you go from outside toward the particle center.

Unlike the electron whose volume of charge is contained in a thin shell. The peak charge volume of the proton (+4/3 Q) fills the outer tier because the particle is physically smaller. The inner tier (-1/3 Q) has a mode shape similar to that of an electron, but instead of its peak density being microscopically greater than 1.0, its peak density is microscopically greater than 2.0. As one progresses from the outside of the proton toward the inside, the surface area of the charge volume decreases. Because of this and

Matter With Electromagnetic Resonance

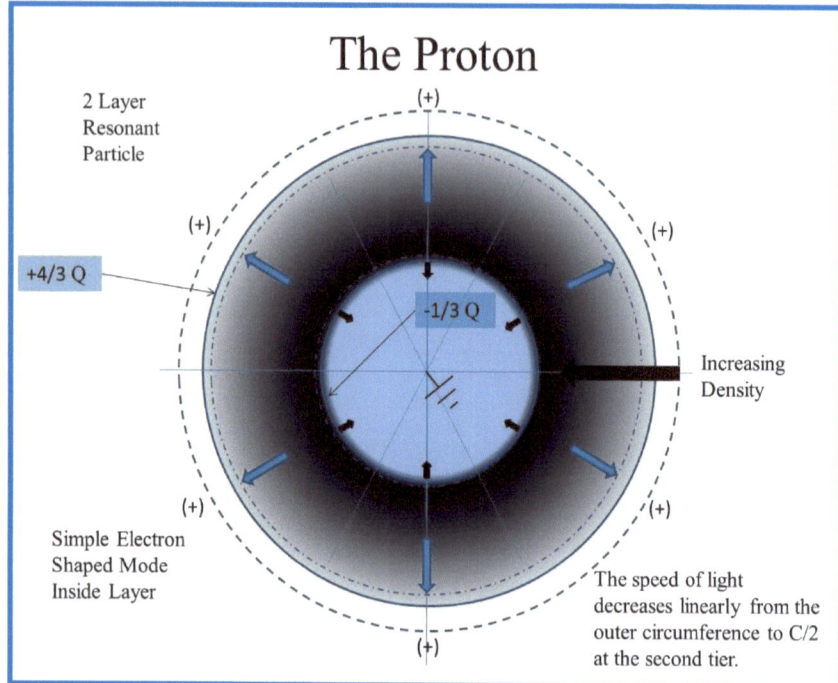

Fig 1.5 The two-tiered resonant cavity structure of the proton allows a positive charge (positive pressure) pointing out to be stable. Halfway into the particle, even though the voltage is ½ that of the outside, that ½ voltage pressure is pushing on only ¼ of the area. The increased pressure at this point in the particle squeezes space so that the peak index of refraction becomes 2.0.

the conservation of energy the volts/meter goes up and the density of the compressed space increases. This density change alters the speed of light linearly such that the wave can rotate smoothly around the central virtual ground. For the two tiers there is a voltage ratio of 2:1, an area ratio of 4:1, and a volume ratio of 8:1. There are seven equal volumes in the outer tier and one in the center. When all of the volumes in the outer tier are polarized outward this stable particle is called the proton. When three of the

Matter With Electromagnetic Resonance

volumes in the outer tier "flip" polarization (which is allowed for electromagnetic modes but is less stable) the net charge of the particle becomes zero and is called the neutron. In my opinion it is likely that when a proton disintegrates, the +4/3 Q of the outer tier is made up of 2 waves. They are spinning in the same direction in one plane while spinning in opposite directions in the other plane forming a standing wave. Upon disintegration each wave has +2/3 Q. These waves are released as well as the central volume of -1/3 Q. It is my belief that these transient waves are what physicists call quarks. This two-tiered higher order mode also explains the magnetic moment discrepancy between the electron and proton. The proton has 4/3 Q at twice the density rotating about its center when compared to an electron. Instead of the electron's moment being ~1820 times as big, these differences would make it about 690 times that of the proton. This new theory estimate is right in line with measured numbers.

As stated earlier, to make a neutron, one simply needs to flip the polarity of three of the seven volumes in the outer tier. This is allowed in electromagnetic modal analysis. The particle is now electrically neutral in the far field. However, it is not as stable as a proton. With the polarity going the opposite direction in three regions, the voltage gradient is different in those regions. Another way of looking at it is that the neutron's oppositely charged cells will tend to bulge out farther from the negatively charged inner tier than the positive cells do. This may explain why the neutron is neutral but has a negative magnetic moment. This slight negative bulging probably also explains why protons can snuggle

Matter With Electromagnetic Resonance

up close to neutrons in an atomic nucleus. The proton's positive regions are attracted at close distances to the neutron's slight negative bulges. From a distance, the neutron appears electrically neutral.

Mass vs. Space Density Function Derivation for Visible Matter (Spin Modes)

Ohm's Law

$V = I \cdot Z$ (Volts = Current x Impedance) **(Eq. 1.4)**
$I = Q \cdot f$ (Current = Charge x frequency or Coulombs per second) **(Eq. 1.5)**
$Z = \eta \cdot Z_0$ (Impedance = Density or index of refraction · free space impedance ($Z_0=1$)) **(Eq. 1.6)**
(This voltage pressure is in series with free space, therefore it adds to the free space voltage. The impedance goes up.)

Ohm's Law Becomes
$V = Q \cdot f \cdot \eta \cdot Z_0$ (where $Z_0=1$, these are all relative densities compared to free space) **(Eq. 1.7)**

At any one frequency (at any radial point inside a particle) Ohm's Law becomes

$V_r = Q_r \cdot \eta_r$ or $V_r/Q_r = \eta_r$ **(Eq. 1.8)**

Even though the amount of charge Q in a single resonant cavity is constant, if you break up this volume charge Q into

Matter With Electromagnetic Resonance

nested surface charges Q_r with constant thicknesses, as you move from the outside to the inside of the resonance, the charge area decreases toward the center. This causes the density to increase.

Energy Equations

$$e = mC^2 = VQ \qquad \text{(Eq. 1.9)}$$

Speed of light changes linearly with the density changes

$$C = C_0/\eta \qquad \text{(Eq. 1.10)}$$

Substitute for C and V (Ohm's Law) and solve for mass(m)

$$m = \eta^2 \cdot V \cdot Q/C_0^2 \qquad \text{(Eq. 1.11a)}$$
$$m = Q^2 \cdot f \cdot \eta^3/C_0^2 \qquad \text{(Eq. 1.11b)}$$
$$m = Q^2 \cdot f \cdot \eta^3 \qquad \text{(Eq. 1.11c)}$$

Voltage (V) changes proportional to the radius (r).

Surface charge (Q_r) changes proportional to the square of the radius (r).

Density (η) changes inversely proportional to the radius (r).

Frequency (f) changes inversely proportional to the radius (r).

At resonance all of these terms have a relationship to the radius of the particle, as stated above. In addition, since all of these are relative densities, C_0 divides out and can be considered = 1.0

Matter With Electromagnetic Resonance

Because of the resonant relationship between an η and f, mass relative to the index of refraction (η space density) becomes

$$m = Q^2 \cdot \eta \cdot \eta^3 \qquad \text{(Eq. 1.11d)}$$

Minimum resonant Volume relative to radius at Maximum Density (see fig. 1.1)

volume = $(r/\eta)^3$ (4/3·Pi drops out because it is constant for all spheres) **(Eq. 1.12)**

Calculate Standard Density for maximum mass at minimum volume

$$\text{mass / volume} = Q^2 \cdot \eta \cdot \eta^3 / (r/\eta)^3 \qquad \text{(Eq. 1.13a)}$$

Because of the resonant relationship between r and η

$$\text{mass / volume} = Q^2 \cdot \eta \cdot \eta^3 \cdot \eta^3 \cdot \eta^3 \qquad \text{(Eq. 1.13b)}$$
$$\text{mass / volume} = Q^2 \cdot \eta^{10} \qquad \text{(Eq. 1.13c)}$$

Because (η) Density of space is only being compressed in a constant amount of space Q (not the entire volume of the particle) the mass ratio and the density ratio are the same value.

$$\text{mass} = Q^2 \cdot \eta^{10} \qquad \text{(Eq. 1.13c)}$$
$$\text{mass}_1 / \text{mass}_0 = Q_1^2 \cdot \eta_1^{10} / Q_0^2 \cdot \eta_0^{10} \text{ where } Q_0^2 \cdot \eta_0^{10} = 1 \quad \text{(Eq. 1.14)}$$
$$\text{mass}_1 = Q_1^2 \cdot \eta_1^{10} \text{ mass}_0 \text{ where mass}_0 = 9.1 \times 10^{-31} \text{ kg (electron mass)} \qquad \text{(Eq. 1.2)}$$

Because of the relationship between mass-energy and resonant frequency-energy

$$\text{Freq}_1 = Q_1^2 \cdot \eta_1^{10} \text{ Freq}_0 \text{ where Freq}_0 = 1.236 \times 10^{20} \text{ Hz (electron frequency)} \qquad \text{(Eq. 1.3)}$$

Matter With Electromagnetic Resonance

Chapter Two
Bounce Modes and the Neutrino

The Self-Sustaining "Bounce Modes"

The lower pictures in Figure A.3 (c,d,e,f,g) comprise cross-sections of what I have called bounce mode particles. These electromagnetic resonant cavity modes also fall out directly from Snell's Law. For each bouncing case to be resonant and self-sustaining, the change in the index of refraction at the particle boundary must be a specific value related to the critical angle. The critical angle is determined by the wave path bouncing inside each sphere. Each path eventually ending up back at the wave's starting point. The critical angle is defined as the angle between the path of the wave and the radial line that intersects that wave at the boundary.

The numerator index of refraction (density of space) is outside (free space $\eta \sim 1.0$) while the denominator index of refraction is just

Matter With Electromagnetic Resonance

> Snell's Law[$\theta_{incident} = \theta_{critical} = \arcsin(\eta_1/\eta_2)$]

inside at the resonant cavity boundary. In addition, the bounce path length around must be an even integer multiple of a wavelength such that when it gets back to the starting point, the wave is in phase with itself. These bounce modes spin slowly because it takes many wavelengths to get back to the starting point. Also, because the voltage pressure is tangential to the boundary instead of normal to it, as in the electron and proton, there is no charge (+ or -) presented to the outside world. Instead, a weak constant magnetic field is present at the boundary. These bounce mode particles have very large masses (high resonant frequency), have no charge, and are extremely small. Approximately 90% of the stable particles that this theory predicts fall into this category. Bounce mode particles are good candidates for Dark Matter. Of the approximately seventy possible bounce mode particles (stable and unstable) predicted by this theory, only about twenty meet the criteria described above for stability. The other fifty particles that tried to form eventually flew apart. Those fifty are good candidates for Dark Energy.

All but one of the particles (the neutrino) that this theory predicts uses Snell's Law for total internal reflection. [I will describe the neutrino at the end of this chapter. It has a different mechanism that sustains it.] The "visible" and "dark matter" particles are different from each other in some important ways. The electron, proton and neutron have a voltage (pressure) gradient parallel to the radius of each resonant sphere and thus present some constant charge, whether positive or negative, to the outside world

Matter With Electromagnetic Resonance

(see Figures 1.3 and 1.5). One could say that their voltages are "in series" with free space because of this radial orientation. The smaller more massive particles that are predicted by the bounce modes have their voltage oriented parallel to the circumference of the sphere (see Figure 2.1). You could say that their voltages are "in parallel" with free space. As stated earlier, these bounce mode particles have a weak magnetic field on their exterior because of the relationship between the electric and magnetic fields. They might be considered weak magnetic monopoles.

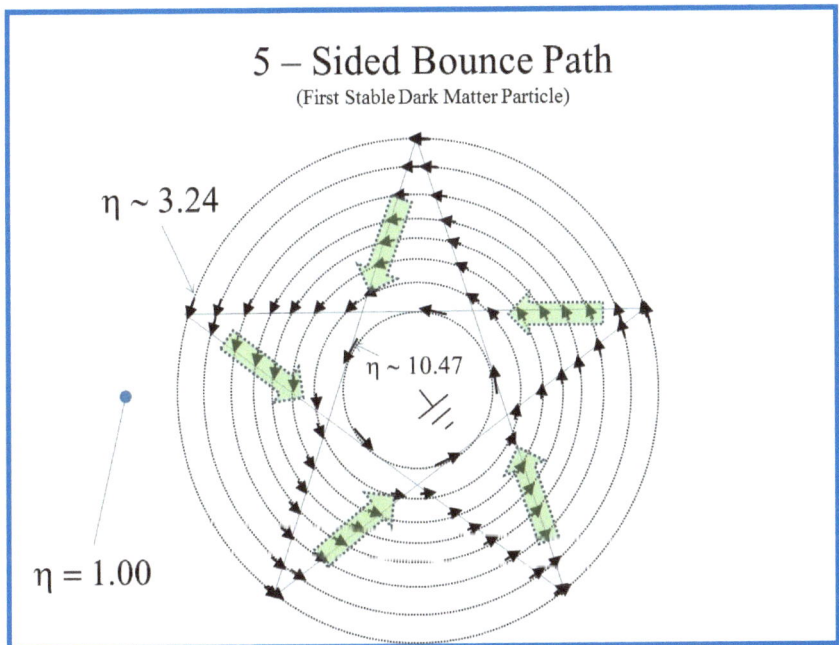

Fig 2.1 The path (shaded large arrows) that a wave takes through the interior of a 5-sided bounce mode particle. The electric field (small dark arrows) is polarized circumferentially. The density of space (index of refraction) increases linearly from the outside to the inner diameter where the density of space is approximately equal to the outer boundary space density squared ($\eta_{inside} \sim \eta_{outside}^2$ or $10.47 \sim 3.24^2$).

Matter With Electromagnetic Resonance

Like visible matter, the density of space increases linearly toward the center of the sphere. In the plane shown in Figure 2.1 the electromagnetic wave propagates through the sphere at a specific critical angle determined by Snell's Law. In the same Figure 2.1, which has a 5-sided electromagnetic path inscribed inside of a sphere (critical angle = 18 degrees), it can be shown that the formula for the critical angle for this "odd number" path is [θ_c = $90_{degrees}/N$] where N is the number of sides. If the number of sides is an even number then the critical angle is [θ_c = $180_{degrees}/N$]. If the wave traveling on the bounce path through the sphere has enough electrical length when it reaches the other side, it will be in phase with the waves that reached the other side taking a circumferential path (see Figure 2.2). These circumferential paths are taken by waves in the other plane. Because of the linearly increasing density toward the center of the sphere, these circumferential waves can rotate smoothly to the other side of the particle. They meet up with that part of the wave travelling close to the center. At resonance, this circumferential path length to the other side is 1.5 wavelengths. (Remember, the circumference for resonance must be three wavelengths in whatever density is present).

Looking at Figure 2.1 again, for the wave taking the bounce path, the lens-like nature of the changing density causes a flip in polarity (180 degrees or ½ wavelength equivalent) as the wave reaches the opposite side. Therefore, to be equal in phase to the circumferential waves there must be enough bounce path length to be an integer multiple of a wavelength. It also appears that the integer number of wavelengths must be even. When the wave

Matter With Electromagnetic Resonance

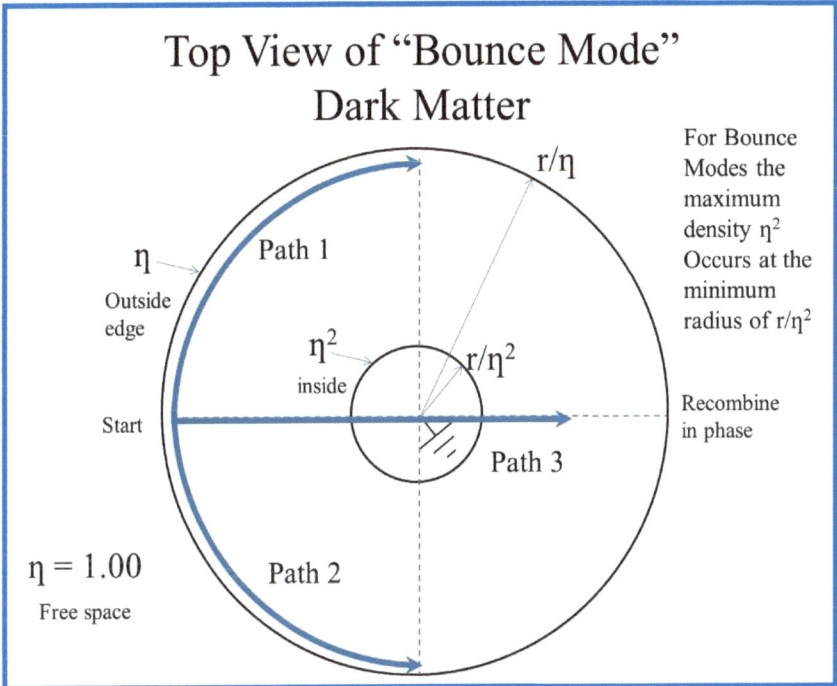

Fig2.2 *Three paths that a wave takes to get to the other side of a bounce mode particle. The space density (index of refraction) increases linearly from the outside to the inner diameter where the space density is approximately equal to the outside density squared ($\eta_{inside} \sim \eta_{outside\ edge}^2$).*

passes through the smallest radius at the densest part of the particle it must be in phase with that circumferential wave as well.

The Density Function for Bounce Modes
(Dark Matter/ Dark Energy)

Another important difference between visible and Dark Matter is that while the volume of compressed space was constant for visible matter (1.0 Q = 0.7 x 10^{-45} meter³), for bounce modes the

Matter With Electromagnetic Resonance

volume Q is actually compressed. This extra shrinkage requires more energy and a different function for the relationship between mass, frequency and index of refraction (η). The new equations are shown below. The derivation is shown on pages 54-56.

$$\text{Mass}_1 = Q_1^2 \cdot \eta_1^{13} \cdot \text{Mass}_0 \text{ where } \text{Mass}_0 = 9.1 \times 10^{-31} \text{ kg}$$
(electron mass) **(Eq. 2.1)**

And because of the relationship between mass-energy and resonant frequency-energy

$$\text{Freq}_1 = Q_1^2 \cdot \eta_1^{13} \cdot \text{Freq}_0 \text{ where } \text{Freq}_0 = 1.236 \times 10^{20} \text{ Hz}$$
(electron frequency) **(Eq. 2.2)**

Equations 2.1 and 2.2 can be used when calculating resonant particle masses with resonant frequencies between the proton frequency (2.25×10^{23} Hz) and the Planck frequency (~1.86×10^{43} Hz).

Using these criteria, the density function and the critical angle from Snell's Law, it is possible to create Table 2.1 to see which frequencies are capable of sustaining stable resonance.

Using these criteria, the number of stable frequencies using a five percent window (roughly the range between a diameter of $3/\pi$ Lambda and 1.0 Lambda) appears to be 20. The different critical angles are shown. To be sure, the straight line calculations of the bounce path lengths through the particles are estimates. One can see from Figure 2.1 that the incremental path length through the denser part of the particle is slightly greater than the path length

Matter With Electromagnetic Resonance

#	Index of refraction η (space density)	# sides	critical angle (deg)	Resonant Frequency (Hz)	Wavelength (meters)	outside diameter (meters)	effective inside diameter (meters)	mass (kg)
1	3.236	5	18.00	5.28E+26	5.69E-19	1.68E-19	5.19E-20	3.88E-24
2	11.474	36	5.00	7.38E+33	4.06E-26	3.38E-27	2.95E-28	5.43E-17
3	15.290	48	3.75	3.08E+35	9.72E-28	6.07E-29	3.97E-30	2.27E-15
4	19.107	60	3.00	5.59E+36	5.36E-29	2.68E-30	1.40E-31	4.12E-14
5	19.744	31	2.90	8.56E+36	3.50E-29	1.69E-30	8.58E-32	6.30E-14
6	23.562	37	2.43	8.53E+37	3.52E-30	1.43E-31	6.05E-33	6.28E-13
7	27.381	43	2.09	6.01E+38	4.99E-31	1.74E-32	6.36E-34	4.42E-12
8	28.017	88	2.05	8.10E+38	3.70E-31	1.26E-32	4.51E-34	5.96E-12
9	31.200	49	1.84	3.28E+39	9.14E-32	2.80E-33	8.97E-35	2.42E-11
10	31.836	100	1.80	4.27E+39	7.03E-32	2.11E-33	6.63E-35	3.14E-11
11	32.473	51	1.76	5.52E+39	5.44E-32	1.60E-33	4.92E-35	4.06E-11
12	35.019	55	1.64	1.47E+40	2.04E-32	5.56E-34	1.59E-35	1.08E-10
13	35.655	112	1.61	1.86E+40	1.61E-32	4.32E-34	1.21E-35	1.37E-10
14	36.292	57	1.58	2.34E+40	1.28E-32	3.37E-34	9.29E-36	1.72E-10
15	39.475	124	1.45	6.98E+40	4.30E-33	1.04E-34	2.63E-36	5.14E-10
16	40.111	63	1.43	8.60E+40	3.49E-33	8.31E-35	2.07E-36	6.33E-10
17	40.748	128	1.41	1.06E+41	2.84E-33	6.66E-35	1.64E-36	7.77E-10
18	43.294	136	1.32	2.32E+41	1.29E-33	2.85E-35	6.59E-37	1.71E-09
19	43.931	69	1.30	2.81E+41	1.07E-33	2.32E-35	5.29E-37	2.07E-09
20	44.567	140	1.29	3.38E+41	8.87E-34	1.90E-35	4.26E-37	2.49E-09

Table 2.1 Space Densities (η…..indices of refraction) that support stable total internal reflection and their associated frequencies for bounce modes. The lightest stable bounce mode particle (5-sided) is ~ 2400 times more massive than a proton as well as being ~ 2400 times smaller than a proton.

near the outside for the same change in radius. However, near the center of the particle, with the virtual ground in close proximity, the space density quickly goes back to 1.0 where the voltages from opposite sides cancel each other out. Because of these near field effects, these path lengths nearest the center are a little shorter than they appear visually. Resonance itself plays a part in the depth of increasing space density (amount of compressed space). The depth will actually modify itself (within a few percent margin). The wave travelling through the interior, which is in phase with the circumferential wave, will be reinforced. What this makes clear to me is that all of the potential bounce mode

Matter With Electromagnetic Resonance

particles will have different likelihoods of being stable. Their half-lifes will be different. The ones closer to ideal will last longer than the others. This variable stability could be what physicists are looking at when they talk about dark matter and dark energy. If there are seventy potential dark matter particles predicted by this theory but only twenty or so fall into the five percent stability range, then all of the others tried to resonate, but ultimately flew apart, expanding space. Even the twenty or so particles that remain today have slightly different stabilities. It is possible that the least stable particles remaining today are now disintegrating, expanding space. These disintegrating particles might be what physicists call Dark Energy.

Compressing Space and Changing from Visible to Dark Matter

Equation 1.3 is the derived density function of space for frequencies up to and including the electron and proton frequencies.

As stated earlier, the net charge of Q is constant in the

$$f_0 \cdot Q_1^{\,2} \cdot \eta_1^{\,10} = f_1 \qquad \text{(Eq. 1.3)}$$

Where $f_0 = 1.236 \times 10^{20}$ Hz (the electron frequency)

η_1 = index of refraction, relative density of space induced by the voltage at the new resonant frequency f_1.

Q_1 = the amount of space being compressed.

Matter With Electromagnetic Resonance

frequency range from the electron to the proton even though the density of space in that compressed volume is changing. For the electron, which has a thin shell shape for Q, the maximum density of space on the inside of the shell is only slightly greater than 1.0. Squeezing Q down, close to the size of the proton outer tier, (remember, the proton and neutron are two-tiered resonant cavities where 2.25×10^{23} Hz is the resonant frequency) increases the thickness of the shell such that the Q=1 almost completely fills the outer tier. At that point in the radius of the proton the density on the inside has increased but has not reached the value of 2.0. At Q = 1.333, right at the inner tier edge ($r_{inner} = .5r_{outer}$), the space density peaks at exactly 2.0. This density is required for a two-tiered resonant cavity to be stable. Upon increasing the pressure and squeezing further, at ~2.6×10^{23} Hz the charge volume of Q = 1 (~ 0.70×10^{-45} m^3) would completely fill the resonant volume. To squeeze Q to an even smaller volume (higher resonant frequency) would require additional pressure (voltage) greater than Equation 1.3. In this region the physical volume of Q is now shrinking by an additional radius^3. Since the resonant voltage (frequency) is proportional to 1/r, it might follow that the frequency and density relationship above 2.6×10^{23} Hz becomes Equation 2.2.

It turns out that equation 2.2 is correct, however, trying not to jump to mathematical conclusions and because of an interesting phenomenon that occurs when you squeeze space beyond 2.6 x

$$f_0 \cdot Q_1^{\,2} \cdot \eta_1^{\,13} = f_1 \qquad \text{(Eq. 2.2)}$$

Matter With Electromagnetic Resonance

10^{23} Hz, it is important to take all of the changes that occur into account before making a pronouncement like Equation 2.2. The most noticeable change that happens is the polarization of the electromagnetic wave. The (+) and (-) charges below 2.6 x 10^{23} Hz had room to exist on both the inside and outside of the volume Q. Volume Q had the shape of a hollow spherical shell. The voltage polarization was radial. Above 2.6 x 10^{23} Hz the volume is so small and virtually filled with compressed space that there is little room for this separation of charge. Now the other polarization, where the electric field is parallel to the circumference of the sphere is the most stable polarization for these bounce modes. Because of this polarization change, the impedance calculation inside the resonant particle is different. The impedance change requires a recalculation of the space density vs. frequency function. In the frequency domain between the proton (2.25 x 10^{23} Hz) and 2.6 x 10^{23} Hz, it is difficult to predict exactly what the space density function is doing using simple hand calculations. Changing the polarization of the field, squeezing a two-tiered structure down to a single tier while possibly releasing energy complicates this function. How the function changes in this frequency region will not be explored here.

Mass/Energy Percentage

If the Big Bang was an impulse function, as its name implies, then the energy available at each potential resonant particle frequency is a constant (the Fourier transform of an impulse in the

Matter With Electromagnetic Resonance

time domain is an equal amplitude function in the frequency domain). This means that the total mass possible at each frequency is a constant. This makes it easy to figure out mass percentages of particles in the universe without having to know how many particles there are of any particular frequency or mass. There are fewer more massive particles, but their total mass is equal to the total mass of each of the other particles. All one needs to know is how many possible frequencies there are to get the mass ratios.

If I just consider stable particles, the sum of 20 stable bounce modes +2 spin modes (electron and proton, (*the neutron and proton are the same particle except for a mode flip in the neutron, see chapter I*) + neutrino (separate mechanism for stability) = 23 particles. It is interesting to note that the percentage of visible particles would then be ~13% (3/23). While the actual percentage of visible vs. dark matter is not universally agreed upon, nowadays it is usually stated to be a low number near ten percent.

Looking at the bigger picture and including the unstable bounce modes as well, the numbers are... 70 bounce modes + 2 spin modes + neutrino = 73 possible density states (73 separate frequencies). Ordinary matter = 3/73 = 4.1%. Dark Matter (stable bounce modes) = 20/73 = 27.4%. Dark Energy (unstable bounce modes) = 50/73 = 68.5%. These numbers by themselves don't prove anything, but they line up nicely with current estimates for these different entities. In addition, it is possible that the least stable of the "stable bounce mode particles" are disintegrating now, which could be contributing to the observed expansion of the universe.

Matter With Electromagnetic Resonance

Dark Neutron

One interesting unstable bounce mode particle tries to form when the number of sides or bounce paths around the inside of the sphere is three. This unstable particle is the lightest possible bounce mode particle and the closest in energy to protons and neutrons. This particle has been named the "Dark Neutron" by me for the following reasons. In a collapsing star, gravity pulls visible matter closer together. If the gravity is strong enough, visible matter will try to be squeezed first into a "dark neutron". Unfortunately the dark neutron is unstable and will attempt to disintegrate unless it is surrounded by extra space density on its exterior. This is not unlike the visible neutron which is unstable unless it is surrounded by protons in a nucleus. Once these dark neutrons are created and try to resonate, most are near the center of the collapsing star. Some are near the outside. It seems likely to me that the ones near the center can be kept from disintegrating because they are surrounded uniformly by all of the other particles and increased pressure. To me this sounds just like a neutron in a nucleus. The dark neutrons near the outside are not so lucky. There is less pressure on the outside of the star. Dark neutrons that are near the edge are more vulnerable. They are more likely to disintegrate and explode. This explosion on the surface will release huge amounts of energy. Extra energy should compress the star even further. This mechanism seems to fit with the observed supernova of stars.

Matter With Electromagnetic Resonance

Neutrino

All of the stable particles predicted by this theory can be thought of as self-sustaining excess space density bubbles. What happens if a space density bubble (1.0 Q) forms and it has the energy of an electron and it is not spinning (rather it is an expanding compressive wave)? Because the wave is not spinning, the bubble cannot sustain itself using spin resonance. When this happens, the density bubble expands until it hits a new limit. Since the volume of compressed space Q is constant, as the surface area of this expanding bubble grows, the bubble wall thickness shrinks. When the wall thickness reaches the Planck length (approximately 1.62×10^{-35} meters) it has to stop. The Planck length is thought to be the smallest possible dimension of space. Here the density bubble is stable because of the conservation of energy. Using the volume of Q that was determined earlier (~7.0×10^{-46} meters^3), when the thickness of this low mass energy density bubble reaches the Planck length, the diameter of the bubble is ~3.7×10^{-6} meters. This would correspond to a resonant frequency of ~7.7×10^{13} Hz. This is right in the middle of the infrared region of the electromagnetic spectrum. The mass energy of such a bubble would be ~ 1.6 million times less than that of an electron. This "hard stop" for the wall thickness is interesting in that it does not require a contained travelling wave to sustain it. However, because the bubble has a specific size, this density bubble can couple to waves at its resonant frequency that happen to be

Matter With Electromagnetic Resonance

traveling by. The predictions and conclusions stated above fit the known characteristics of the neutrino. The calculated energy level is very close to the published energy level (Table 2.2 reference A). The fact that it can couple to infrared radiation impinging upon it might possibly explain the changing flavor of a neutrino as it travels through space. When infrared waves of different polarizations pass through the bubble, it is possible for the wave to briefly couple to the bubble until another wave comes along and knocks it out.

New Theory Supported by Measured Data

Some interesting galactic and extra-galactic Gamma Ray data has been published and are cited in references C through F of Table 2.2. These measurements support the frequency and energy combinations predicted by this theory. Even the Higgs Particle energy that was measured in a collider in the spring of 2013, falls very close to the frequency/energy spot where an unstable bounce mode particle is predicted (Table 2.2 references E, F). Some very high energy gamma rays measured recently line up nicely with the first stable 5-sided bounce mode dark matter particle predicted by this theory. Unstable 8-sided and 12-sided dark matter (dark energy) energies straddle the 5-sided stable particle (table 2.2 references D, F). The energy level predicted by the quasi-stable Dark Neutron also shows up in measured Gamma Ray data from space (Table 2.2 references C, D, F). One of the surprising features of the measured Gamma Ray data that I have looked at is

Matter With Electromagnetic Resonance

Measured Data from the Galaxy and beyond Support this Resonant Theory of Matter

Particle	η Space Density	Resonant Frequency Hz	Energy electron volts	Relative Mass vs. proton	mass kg	Reference Measured Data to support
Neutrino	η = 2.25	7.665E+13	3.174E-01	3.406E-10	5.643E-37	A
Electron	η = 1	1.236E+20	5.119E+05	0.00055	9.100E-31	B
Proton	η = 2	2.250E+23	9.318E+08	1.00000	1.657E-27	C D F
Dark Neutron #sides = 3	η = 2	1.013E+24	4.193E+09	4.5	7.454E-27	C D F
Unstable Higgs or Galactic source #sides = 8	η = 2.613	3.274E+25	1.356E+11	145.52	2.411E-25	E F
First Stable Dark Matter #sides = 5	η = 3.236	5.275E+26	2.185E+12	2344.52	3.884E-24	D F
Unstable Dark Matter #sides = 12	η = 3.864	5.285E+27	2.189E+13	23489.50	3.891E-23	D F

Table 2.2 *Predicted Particles (stable and unstable), their resonant frequencies, energies and space densities. All of these particles except the neutrino have corresponding measured Galactic Gamma Ray data "bumps". References C through F show measured Gamma Ray data that line up with the last five particles in this table. The neutrino energy/mass was gleaned from different measured astronomical data.*

that the high energy bumps predicted by this theory appear to be the only bumps in the data. I find this very interesting and supportive. This data does not prove the theory, but they are additional indicators of its validity and have spurred me to look further. Table 2.2 shows the location of these predictions in the frequency domain as well as some other particles like the neutrino. The mass/energy of a neutrino being less than one millionth that of an electron is easily predicted using this theory and constant volume Q. (Table 2.2 references A^2, B^3, C^4, D^5, E^6, F^7).

Matter With Electromagnetic Resonance

Mass vs. Space Density Function Derivation for Dark Matter (Bounce Modes)

> ***Ohm's Law***
> $V = I \cdot Z$ (Volts = Current · Impedance) (Eq. 1.4)
> $I = Q \cdot f$ (Current = Charge · frequency or Coulombs per second) (Eq. 1.5)
> $Z = Z_0 / \eta$ (Impedance (goes down) = Free space impedance ($Z_0=1$) / Density or index of refraction) (Eq. 2.3)
> *(The voltage pressure for bounce modes is in parallel with free space therefore the voltage inside is equal to the free space voltage. However, inside the particle the pressure (volts/meter) goes up because the radius is shrinking.)*
>
> ***Ohm's Law becomes***
> $V = Q \cdot f \cdot Z_0 / \eta$ (where $Z_0=1$, these are all relative densities compared to free space) (Eq. 2.4)
> *See Figure 2.1 - The voltage pressure is parallel to the circumference*
>
> *For one frequency (at any radial point inside a particle)*
> ***Ohm's Law becomes***
> $V_c = Q_c/\eta_c$ or $Q_c/V_c = \eta_c$ (subscript "c" is circumferential) (Eq. 2.5)

Because these voltages are in parallel;
- Voltage (V_c) is constant
- Charge (Q_c) changes inversely proportional to the radius (r)
- Density (η_c) changes inversely proportional to the radius (r)
- Frequency (f) changes inversely proportional to the radius (r)

Matter With Electromagnetic Resonance

Energy equations
$$e = mC^2 = VQ \quad \text{(Eq. 1.9)}$$

Speed of light changes as the density changes (Bounce mode max density inside = η^2)
$$C = C_0 / (\eta^2) \quad \text{(Eq. 2.6)}$$

Substitute for C and V and solve for mass (m)
$$m = (\eta^2)^2 \cdot V \cdot Q/C_0^2 \quad \text{(Eq. 2.7a)}$$
$$m = Q^2 \cdot f \cdot \eta^4/C_0^2 / \eta \quad \text{(Eq. 2.7b)}$$
$$m = Q^2 \cdot f \cdot \eta^3 \quad \text{(Eq. 2.7c)}$$

At resonance all of these terms have a relationship to the radius of the particle, as stated above. In addition, since all of these are relative densities, C_0 divides out and can be considered = 1.0

Because of the resonant relationship between f and η
Mass relative to the index of refraction (η space density) becomes
$$m = Q^2 \cdot \eta^4 \quad \text{(Eq. 2.7d)}$$

Minimum resonant Volume relative to radius at Maximum Density (see Figure 2.1)
volume = $(r / \eta^2)^3$ (4/3 · pi drops out because it is constant for all spheres) **(Eq. 2.8)**

Calculate Standard Density for maximum mass at minimum volume
$$\text{mass / volume} = Q^2 \cdot \eta^4/(r / \eta^2)^3 \quad \text{(Eq. 2.9a)}$$

Because of the resonant relationship between r and η
$$\text{mass / volume} = Q^2 \cdot \eta^4 \cdot \eta^3 \cdot \eta^6 \quad \text{(Eq. 2.9b)}$$
$$\text{mass / volume} = Q^2 \cdot \eta^{13} \quad \text{(Eq. 2.9c)}$$

Matter With Electromagnetic Resonance

Because (η) Density is still only compressing a constant amount of space Q (albeit a smaller volume) the mass ratio and the density ratio are the same value.

mass = $Q^2 \cdot \eta^{13}$ (Eq. 2.10)

$mass_1/mass_0 = Q_1^2 \cdot \eta_1^{13} / Q_0^2 \cdot \eta_0^{10}$ where $Q_0^2 \cdot \eta_0^{10} = 1$ (Eq. 2.11a)

$mass_1 = Q_1^2 \cdot \eta_1^{13}$ $mass_0$ where $Mass_0 = 9.1 \times 10^{-31}$ kg (electron mass) (Eq. 2.1)

Because of the relationship between mass energy and resonant frequency energy

$Freq_1 = Q_1^2 \cdot \eta_1^{13} \cdot Freq_0$ where $Freq_0 = 1.236 \times 10^{20}$ Hz (electron frequency) (Eq. 2.2)

Figure 2.3 shows a graph of a modified Equation 2.2 [$f_1 = \eta_1^{13} \cdot f_0$]. This is the space density (index of refraction) function for bounce modes. The amount of compressed space for these cases is close to 1.0 Q. Therefore, $Q_1^2 \sim 1.0$ and has been omitted from the graph. None of these particles is a two-tiered resonant cavity. This is the frequency region where the "bounce modes" live. These particles are very small and have very large masses.

Summary

If one thinks about space (the Aether) as a material with elastic and inelastic properties in combination with Snell's Law, which can be used to predict the boundary behavior of excess space density and voltage pressure waves in that elastic material, one

Matter With Electromagnetic Resonance

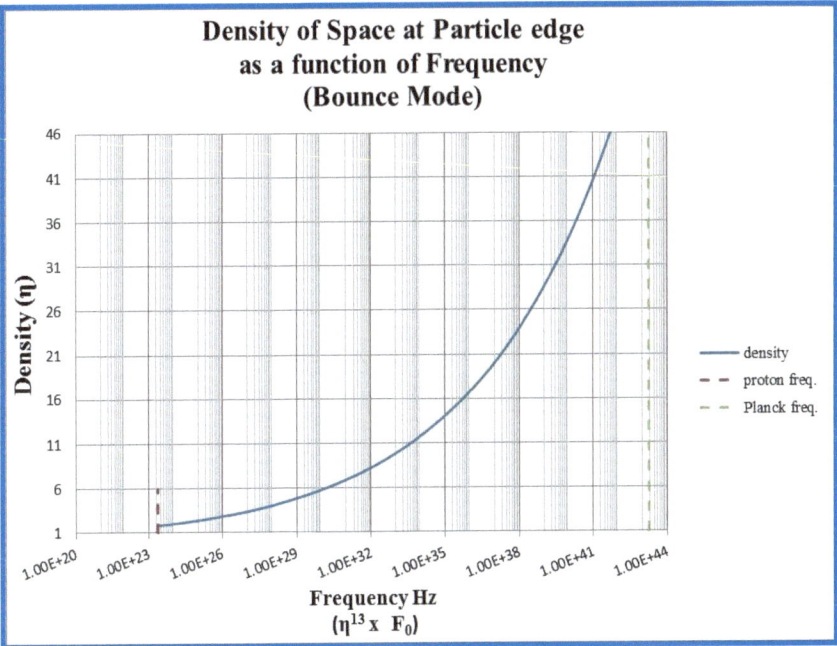

Figure 2.3 *Bounce mode space density function vs. frequency. Q^2 has been left out of this graph because the amount of compressed space Q is very close to 1.0.*

can make a scientific leap to a simple conclusion. All stable particles of visible matter and even dark matter are simply electromagnetic pressure waves trapped inside spherical resonant cavities of their own making. They literally are bubbles of excess space (Aether) with different sizes, densities and wall thicknesses. Even dark energy turns out to be simply unstable dark matter particles that have disintegrated. Neutrinos are also space density bubbles that are trapped when their wall thickness hits the Plank length limit.

Matter With Electromagnetic Resonance

Chapter Three
What do Atoms Look Like?

Let's recap some of the important features of visible matter particles based on this new theory of matter. Theory: Matter comprises spherical resonant electromagnetic cavity waves where the cavities themselves are sustained by the voltage pressure of the wave contained inside. The mechanism of containment is Snell's Law as it pertains to the total internal reflection of an electromagnetic wave at the boundary between two materials of different densities. Space is elastic and compressible (albeit at extremely high voltage pressures).

Matter consists of density bubbles of different sizes, masses and charges.

The density of space is ($\eta_0 \sim 1.0+1.0 \times 10^{-10}$) at the resonant frequency (F_0) of the electron (1.236×10^{20} Hz).

The relative density function of space for visible matter as it relates to frequency is $F_1 = Q_1^2 \cdot \eta_1^{10} \cdot F_0$ where $F_0 = 1.236 \times 10^{20}$

Matter With Electromagnetic Resonance

Hz η_1 = Density at the new frequency Q_1 = the volume of compressed space.

The proton is 1820 times smaller than a free electron.

The proton is 1820 times more massive than a free electron.

The proton and neutron are similar two-tiered resonant particles that differ in the polarization of the voltage (direction of pressure) in their second tiers.

Now that I have an idea about the nature of electrons, protons and neutrons, which includes details about their size, shape, direction of pressure (+/- charge Q) as well as their resonant frequencies, I will explore how they might fit together to form atoms. I will tell you right now that the actual configuration of atoms is both surprising and simple and is quite a bit different than what is accepted today. This new theory when applied to atomic sizes fits the measured values for atoms very closely.

Imagine free protons, neutrons and electrons bouncing around in the early universe. There is a natural arrangement of these particles based solely on their sizes, distribution of charge and the density of space. Protons will bounce off of other protons. Free electrons will bounce off of other free electrons. Free neutrons are unstable alone but because of the arrangement of plus (+) and minus (-) charged cells on their exterior, they can attract one another as well as protons. Remember, protons and neutrons are nearly the same size but their second tier resonant cavity cells have different values (+/-) of charge in them. Electrons are ~1820 times larger than protons and have the opposite sign charge (-) on their outside. However, because the electron is an extremely thin shelled

Matter With Electromagnetic Resonance

sphere, it has room for a positive pressure (positive charge) on its inside. The direction of positive pressure is toward the virtual ground at the electron's center and normally cancels. What happens when a positively charged proton is attracted to a negatively charged electron that is 1820 times bigger? It seems likely to me that because of its momentum, the tiny proton will actually penetrate the shell of the electron. They are at such different resonant frequencies that their AC (alternating current) resonances don't really interact, but their residual DC (direct current) like charges do. Once inside the electron shell, the positive charge on the outside of the proton sees the positive charge on the inside of the electron and they repel. The proton centers itself to equalize the charge. There is now too much positive charge ($2 \cdot Q$) inside of the electron and it must expand to equalize the density of space back to a net of 1.0 Q. I will discuss the value of the expansion (it is significant) and how to calculate the atomic radii on pages 63-68.

The first stable atom is hydrogen. To construct the remaining elements, the order of the inclusion of additional protons, neutrons and electrons matters. For instance, since protons repel protons, before additional protons can be attracted to a nucleus they must pair up with free neutrons otherwise they would see the positive charge in the nucleus and bounce off. These additional protons need to be included in the center nucleus before additional electrons can be attracted. If not, the total net far field charge is zero and there is no attraction. Because the expanded atomic electrons (lower resonant frequency) are bigger than free electrons, their charge densities are significantly lower at their

Matter With Electromagnetic Resonance

surfaces. New free electrons, being physically smaller than the atomic electrons, can penetrate the outer atomic electron spheres and engulf the nucleus. These new electrons expand just like the first free electrons did. Atoms get filled from the inside out.

Stable Atomic Shells

When constructing stable atomic shells, there are a few rules of stability that must be followed. Since every electron is an unbelievably thin walled spherical shell, surface area and surface charge density are the prime drivers that dictate each shell's radius. Stability also dictates the allowable number of electrons in each shell. Volume stability also plays an important role when constructing the elements.

Because all of the particles are resonant and the resonant frequency dictates a particle's radius, the relative spin (relative rotational frequency) between two particles that are interacting plays a role in the actual physical radius that a stable shell can occupy. It is this relative spin frequency that allows two sets of stable electron shell radii to be formed around a growing atomic nucleus. The closer electron shell is rotating counter to the rotation of the nucleus which gives it a relatively higher frequency. The farther electron shell rotates in the same direction as the nuclear center which gives it a relatively lower frequency. This relative frequency is not unlike the Doppler frequency shift caused by an approaching or receding train. The higher frequency has a smaller radius, the lower frequency has a larger radius.

Matter With Electromagnetic Resonance

Two Electrons for Every Radius

It may seem counter to logic that each allowable atomic electron radius can be occupied by two electrons. However, if they are spinning in opposite directions they can form a standing wave at the same frequency and both occupy the same radius. At that point, the radius is filled and any additional allowable electrons in that shell must be at a microscopically different radius nested up against the first electron pair. Based on the new lower resonant frequency for electrons that have engulfed protons, each completely filled atomic electron shell has a thickness of approximately 4×10^{-26} meters. This dimension is 11 orders of magnitude smaller than the diameter of a proton. Nested electrons in a shell are extremely close to one another.

Figure 3.1 shows a cross section through a notional sphere that contains several completely filled stable nested electron shells. For atoms, stability means not falling apart absent some external force. The actual number of stable configurations for electrons nested around a nucleus is quite large. However, some configurations are more stable than others such as when electron shells are completely filled. The noble gases, helium and neon fall into this category. Other elements with partially filled shells are stable but reactive. Left alone, reactive elements won't fall apart, but in the presence of other partially filled shells on other atoms they couple to form molecules. Molecules consist of adjacent spheres not nested spheres. Molecular coupling will not be discussed here.

Matter With Electromagnetic Resonance

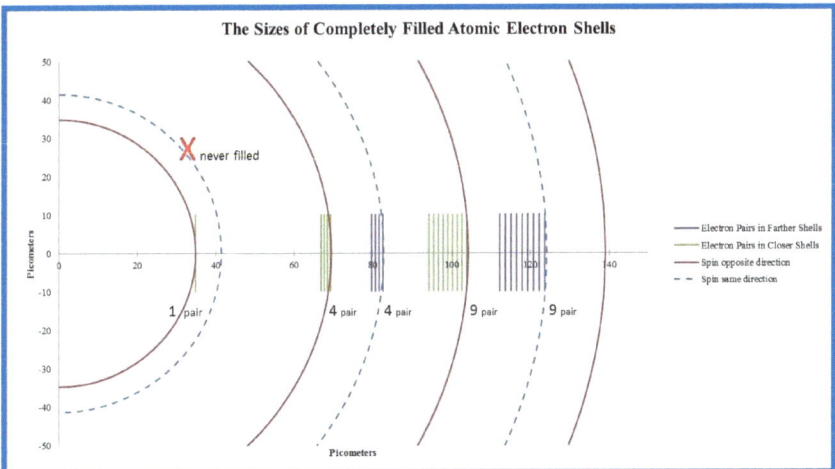

Figure 3.1 *These are the radial locations in picometers (10^{-12} meters) for the first seven completely filled stable electron shells around a spinning nucleus. Volume stability constraints prevent the second lowest shell that would contain another pair of electrons from ever being filled. Note* the graphed spacing for electron pairs is not shown to scale.*

Construct the Elements

There are a number of ways to describe how the elements are assembled. I have decided to write a step-by-step list from the lightest to the heaviest.

A single proton (or proton + neutron) is engulfed by a free electron (opposites attract).

Before capture, the free electron was 1820 times larger than a proton.

After captured, the electron's radius expands by an additional factor of ~ 42 (hydrogen).

An additional proton/neutron pair can penetrate this single

Matter With Electromagnetic Resonance

atomic electron shell. The new pair is pushed to the center where it can attach to the first pair (neutrons have positive and negative charged cells on their exterior that facilitate this attachment). The new pair's momentum, whether linear or angular, is conserved by the spinning nucleus.

Another free electron is now attracted to this excess positive charge. The free electron penetrates the ~ 42 times larger lower density atomic electron and engulfs the nucleus. This newly captured free electron expands just like the first. The first electron contracts a bit to meet it.

Now the first complete stable shell is filled with two oppositely spinning electrons surrounding the 2-protons and 2-neutrons (helium).

During the formation of helium, the angular momentum of the spinning nucleus (previously spinning isolated protons etc.) is conserved (slows down by a factor of 12, $(1/mr^2)$ where m = Mass = 4, and r = nuclear radii ratio = $\cos(30)*2 = r_{helium\ nucleus} / r_{proton}$, and $r_{proton} = r_{electron}/1820$) see figure 3.2

The now slower spinning nuclear center creates two allowable electron shells at different radii.

The smallest stable radius is rotating in the opposite direction of the center (since the center has two protons of opposite spin, each allowable shell contains two electrons of opposite spin).

Space does not expand when protons and neutrons sit side by side.

Electrons have (+) charge on their inside and (−) charge on their outside so concentric electron spheres can nest together without expanding space.

Matter With Electromagnetic Resonance

However, the inside pressure of the electrons pushes against the outward pressure of the protons.

I can calculate the radius of the first completely filled electron shell using the space density relationships developed in this book.

The electron's surface area must increase such that the space density is decreased by a factor of two. Using the new theory's relationship between frequency and space density ($\eta_1^{10} \cdot f_0 = f_1$ where $f_0 = 1.236 \cdot 10^{20}$ Hz and $\eta = 1/\text{sqrt}(2)$), I can calculate the resonant frequency of the fundamental stable atomic electron shell. $(.707)^{10} \cdot f_0 = f_{\text{atomic electron}}$. The atomic electron pair's radius grows by a factor of ~32 to $37 \cdot 10^{-12}$ meters. This nominal radius is modified somewhat by the slower spinning nucleus.

The different relative frequencies of rotation for the electrons, one faster and one slower than the spinning nucleus, change the relative voltage of each stable shell. The voltage pressure in each shell is pushing back against space trying to expand. The closest shell is filled first. It has a higher relative frequency and more voltage. Electrons in this shell push back harder to a slightly smaller radius. (Closest stable atomic electron pair radius = $34.7 \cdot 10^{-12}$ meters). This is the atomic radius of helium.

The addition of the next proton/neutron/electron for lithium slows the nuclear rotation down a little bit more such that the next allowable shell (the shell rotating in the same direction as the nucleus) would have been a little bit larger, $37 \cdot 10^{-12} \cdot 1.068$ or $\sim 41.4 \cdot 10^{-12}$ meters. Volume stability dictates that this shell will never be filled. But this shell's larger harmonic multiples will be filled.

Taking a step back from helium, the hydrogen radius is larger

Matter With Electromagnetic Resonance

than a fully filled shell by the square root of 2 (~49*10^{-12} meters). Shells that are not completely filled are larger than fully filled shells by the ratio of the square root of the number of electrons in a full shell divided by the square root of the number of electrons in the shell. The reason for this is that while the sum of the voltages of the nested electrons add to push back and shrink the radius, this shrinkage is offset (pushed out) by the square root of the number electrons present. (Increasing density pushes out, increasing surface area).

Once the radii for the two possible lowest order 2-electron shells are set, the further slowing of the nucleus due to additional protons and neutrons has a negligible effect. This is because the first two atomic electron's resonant frequencies are now so low (after an expansion of their radii) that a (+/-) rotational frequency change in the nucleus has only a tiny effect.

Subsequent electron shell's radii are now dominated by the first shells radii. In effect, the distance relationship (voltage-pressure) required to reduce the average density of space back to 1.0 between engulfed protons and electrons has been set. For stability, all of the larger radii of completely filled shells are integer multiples of the first two (see figure 3.1).

For heavier elements I use surface and volume stability to predict where additional electrons will go. Counting the number of integer equal areas contained in each layer of a set of concentric spheres (with radii = 1, 2, 3, 4,...) I get (areas = 1, 4, 9, 16,...). Counting the number of additional equal volumes (only volumes added per layer) for the same set of spheres, there are

Matter With Electromagnetic Resonance

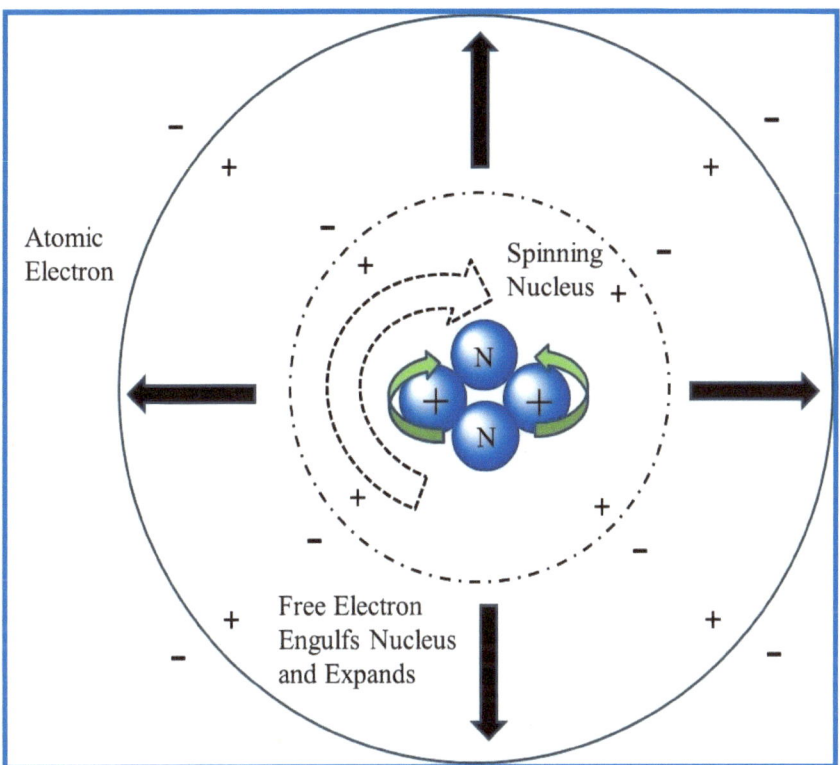

Figure 3.2 *Spinning nucleus is engulfed by free electrons which are 1820 times larger. The plus charge (+) on the inside of the electrons is repelled by the plus charge (+) on the outside of the protons. The two electrons radii expand by approximately thirty-two times to equalize the density of space back to $\eta=1.0$. The spin of the nucleus allows for the formation of two stable sets of electron shells at different radii.*

(added volumes = 1, 7, 19, 37,...). By inspection, even though there are two sets of allowable shell radii based on the spinning nucleus (number of stable electron pairs per shell would have been 1, 1, 4, 4, 9, 9, 16, 16,...), there is not enough stable volume close to the nucleus for the second pair of slower rotating electrons to fit. The next shell that will fit has four pair that rotate

Matter With Electromagnetic Resonance

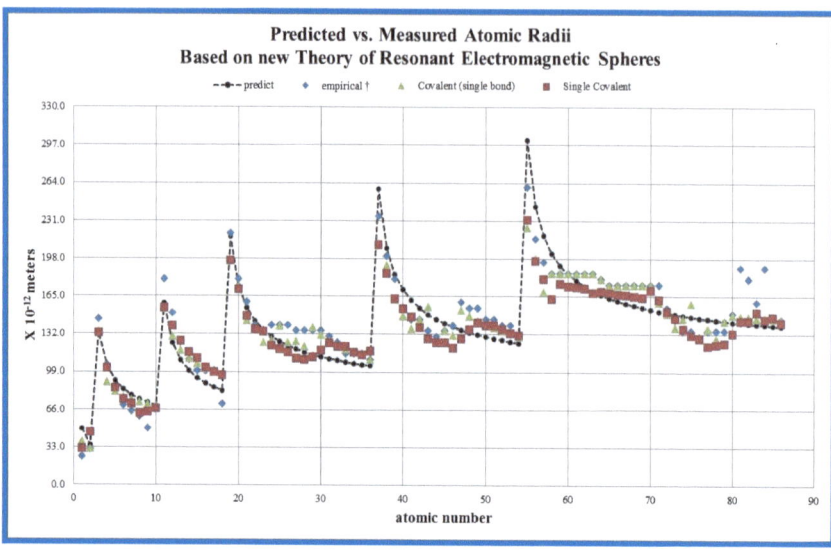

Figure 3.3 *New theory accurately predicts atomic radii.*[8]

in the same direction as the first pair. When the third shell shows up, there is enough available stable volume to fit another set of four slower spinning electrons.

Once you get to the third shell and beyond, volume stability has plenty of room for all of the extra electrons. This is why when you count the number of electron pairs in each filled shell there are 2-fours 2-nines 2-sixteens etc. but only 1-one.

Figure 3.3 shows the calculated radii of all of the elements up to an atomic number of 86 using the new theory. These predictions compare favorably with three sets of measured atomic radii that are also shown in the graph.

Side note: Recently there have been a number of published photos of atomic lattices for various elements. Some of these photographs have been taken using electron microscopes. Every

Matter With Electromagnetic Resonance

photo shows a set of beautifully spherical structures nestled side by side. Sampling theory would seem to suggest that this new theory of atomic structure is correct. To paraphrase, Sampling Theory states that to accurately depict any structure or function, the function must be sampled at a frequency that is at least twice that of the highest frequency component. Otherwise, everything just blurs together. If electrons are orbiting around the nucleus with their diameters unchanged from free electrons, or simply existing in a statistical reality around the nucleus, which is the present day picture, sampling theory suggests that you would never be able to take those beautiful photographs using electrons themselves. If, on the other hand, atomic electrons expand to at least forty times the radius of free electrons as this theory proposes, that would be a sampling ratio of > 40:1. This resolution is more than adequate to take the nice spherical pictures that have been published.

Chapter Four
The Nature of Galaxies and the Unification of Gravity, Electrostatic, Weak, and Strong Forces

Where do Galaxies Come From?

Let's examine the process of shrinking matter by compressing space with the application of more and more voltage pressure. Starting with the lightest stable spin resonant particle, the electron, and continuing past the smaller proton... past the ever shrinking stable dark matter particles, there appears to be an arbitrary stopping point, the Planck length. The Planck length is purported to be the smallest possible physical length. The Planck length is ~ 1.56×10^{-35} meters long. Even though this new theory does not predict a three dimensional stable particle the size of the

Matter With Electromagnetic Resonance

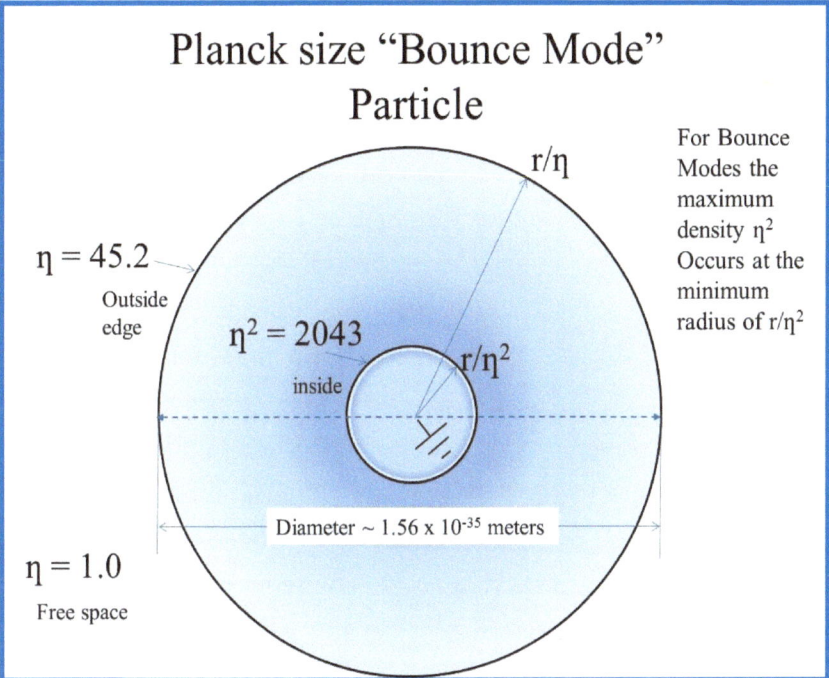

Figure 4.1 *Density of Space ($\eta_{outside}$ = 45.2) needed to make a graded density resonant particle the approximate size of the Planck length. It seems to me that near the center, the space density might be close to its maximum compressibility (η_{inside} = 2043).*

Planck length, the equations provided by the theory can be used to explore the relationship between mass-energy, frequency and space density for something that small. To compress a graded density particle down to that size using this theory would require a space density of $\eta_{outside}$ = 45.2 on the outside edge of the particle and a peak space density on the inside of $(45.2)^2 = \eta_{inside} = 2043$. See Figure 4.1.

Moving from the outside to the inside, the density increases inversely proportional to the radius from 45.2 to 2043. If this

Matter With Electromagnetic Resonance

Planck volume is truly a minimum, it seems logical to me that the value of the compressibility of space may have just about reached a maximum ($\eta \sim 2043$). If that is true, the next question might be how much mass-energy can be compressed into that volume by filling it up to its uniform maximum density (instead of a 1/radius gradient)? Another logical question might be, what voltage (frequency) would it take to achieve that maximum density? Using this new theory, the mathematical tools are now available to make estimates such as these.

Without a doubt it will take more pressure to fill the Planck volume to a uniform density than to a graded density. Because increasing voltage pressure is directly related to frequency, whatever new frequency/voltage is required, each one of these resonant high frequency uniform density volumes will be much smaller than the Planck volume. In reality, the compressed space contained in the Planck volume will simply be vibrating at a higher frequency. Thinking about the new frequency as smaller volumes allows a simple calculation of the total energy contained in the Planck volume. Much like the thin shell of the electron, these tiny volumes can be nested next to one another to fill up the Planck volume. Once an estimate is made for the density (η_{max}) required for uniform filling, the maximum frequency can be found using the new Equation 4.1.

Knowing the maximum frequency, calculations can be made for the quantity of these small volumes inside the Planck sized sphere.

$$f_0 \cdot \eta_{max}^{10} = f_{max} \quad \text{(Eq. 4.1)}$$
Where $f_0 = 1.236 \times 10^{20}$ Hz

Matter With Electromagnetic Resonance

$$N_{volumes} = (f_{max}/f_{Planck})^3 \quad \text{(Eq. 4.2)}$$
Where $f_{Planck} = 1.84 \times 10^{43}$ Hz

Referring to Figure 4.2, estimates can be made about how much the density of space must increase to fill the Planck volume uniformly. Remember, every formula described so far is for a linear change in density. To get uniform density requires twice the voltage. This can be seen in the left side of Figure 4.2. Using a variation of Equation 4.1, the density increase to get twice the voltage is $2^{.1} = 1.0718$. On the right side of Figure 4.2, I have taken a bold step and decided that the original uniform volume of space was a cube and that all of these adjacent cubes were side by side. Starting with a cube gets rid of any voids between adjacent spheres. Voids would make the density non-uniform. Turning a spherical volume into a cube requires ~ 1.2345 times the voltage. Again, with the modified equation 4.1, the density increase works out to be $1.2345^{.1} = 1.0213$. The outside starting density for maximum compression is shown in Equation 4.3.

$$\eta_{start} = 45.2 \cdot 1.0718 \cdot 1.0213 = 49.48 \quad \text{(Eq. 4.3)}$$

therefore the maximum density for uniform filling is shown in Equation 4.4
$$\eta_{max} = \eta_{start}^2 = 2448 \quad \text{(Eq. 4.4)}$$

Solving Eq. 4.1
$$f_{max} = f_o \cdot \eta_{max}^{10} = 9.55 \times 10^{53} \text{ Hz} \quad \text{(Eq. 4.1)}$$

Solving Eq. 4.2
$$N_{volumes} = (f_{max}/f_{Planck})^3 = 1.4 \times 10^{32} \text{ volumes} \quad \text{(Eq. 4.2)}$$

Matter With Electromagnetic Resonance

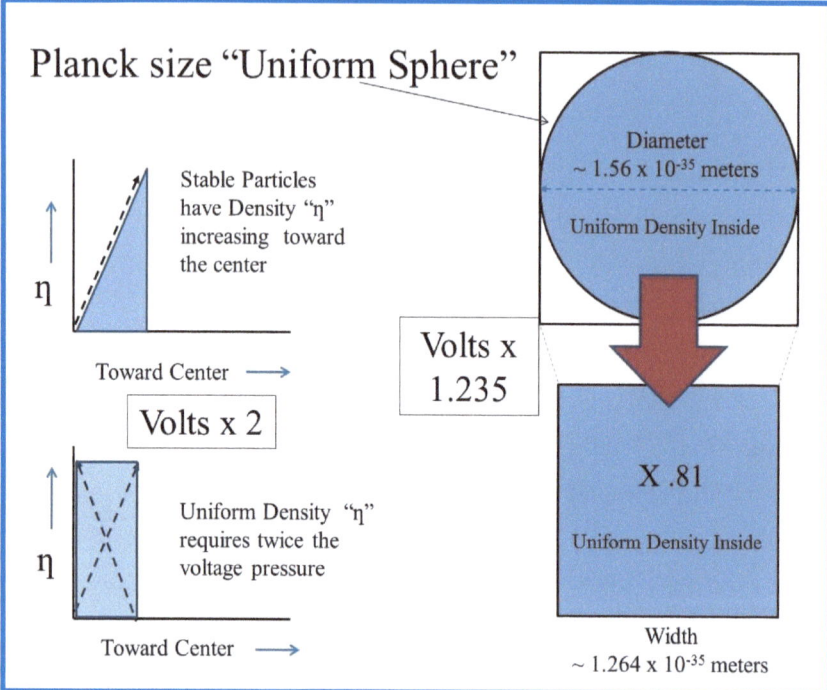

Figure 4.2 *Density of space estimate using the voltage needed to uniformly fill a cubic volume the approximate size of the Planck length. The maximum compressibility of space based on this estimate is (η_{max} = 2448).*

Armed with this information, estimates can be made for how much mass-energy this uniformly filled Planck sized cube can hold.

[$h = 6.626 \times 10^{-34}$ meter² kg sec⁻¹, $C_0 = 3 \times 10^8$ meters sec⁻¹]

Energy for one volume of f_{max}.

$E_{fmax} = m_{fmax} \cdot C^2 = h \cdot f_{max}$ **(Eq. 4.5)**
$C = C_0 \div \eta_{max}$ **(Eq. 4.6)**

Matter With Electromagnetic Resonance

Mass-Energy for one volume of f_{max}.

$m_{fmax} = [\eta_{max}^2 \cdot h \cdot f_{max} \div C_0^2]$ (Eq. 4.7)
$N_{volumes} = (f_{max}/f_{Planck})^3 = 1.4 \times 10^{32}$ (Eq. 4.2)

Total Mass-Energy in the volume of one Planck cube.
$m_{total} = N_{volumes} \cdot m_{fmax}$ (Eq. 4.8)

Solving Eq. 4.8
$m_{total} = 5.899 \times 10^{42}$ kg (Eq. 4.8)

Many readers can see where I am headed here. This is almost exactly the total mass-energy contained in an average galaxy. In fact, I believe that each Planck cubic volume that sat side by side with adjacent Planck cubes gave rise to each galaxy when it expanded during the Big Bang. The resonant particles precipitated out like rain drops as the density of space dropped during expansion. The heaviest (dark matter) particles precipitated out first and to some extent got left behind because of their masses. I believe this is why surveys of the likely location of dark matter look somewhat like spherical halos surrounding galactic centers. I believe this is why astronomers seem to find black holes at the center of most galaxies. [This will be explored in more detail in chapter V]. All of the matter-energy in the universe, visible and dark, as well as dark energy (dark matter particles that tried to resonate but could not) can now be shown to be present in a three dimensional array of cubic Planck volumes sitting side by side prior to the Big Bang.

Using the Planck-Galaxy / mass-energy calculation above, it

Matter With Electromagnetic Resonance

is relatively easy to compare the predicted visible mass in the universe with estimates from other sources.

> Estimate for the number of galaxies in the visible universe.
> $(Radius_{universe} / Radius_{galaxy})^3 = $ Number of Galaxies [9]
> **(Eq. 4.9)**
>
> Solving eq. 4.9
> $(1.306 \times 10^{26}$ meters $/ 1.54 \times 10^{22}$ meters$)^3 = 6.1 \times 10^{11}$ Galaxies **(Eq. 4.9)**
>
> New Theory predicts mass percentages (very similar to estimates from other sources).
> Visible Matter = 3/73 = 4.1% Dark Matter = 20/73 = 27.4% Dark Energy = 50/73 = 68.5%
>
> New theory prediction for the amount of visible matter.
> Number of Galaxies · Total Mass-Energy per Galaxy · 4.1% Visible **(Eq. 4.10)**
> $6.1 \times 10^{11} \cdot 5.899 \times 10^{42}$ kg $\cdot 0.041 = \mathbf{1.475 \times 10^{53}}$ **kg.** **(Eq. 4.10)**
>
> Average of several published estimates (using other techniques) for visible matter in the universe.
> **1.45×10^{53} kg.** [10]

I believe that the expansion of the universe, up to this point in time, can be calculated using the average galaxy starting size being equal to the Planck length. This expansion is shown in equation 4.11. Applying that expansion value in reverse to the background microwave radiation peak frequency of 1.6×10^{11} Hz allows a calculation of the total energy in the total universe (not just the visible). Knowing the total energy in the beginning

Matter With Electromagnetic Resonance

> Expansion of the universe = [average galaxy radius ÷ Planck radius]
> $2.0 \times 10^{57} = 1.54 \times 10^{22}$ meters ÷ 0.75×10^{-35} meters **(Eq. 4.11)**
> Frequency$_{\text{start of universe}}$ = Microwave Background Frequency · Expansion
> 3.2×10^{68} Hz = 1.6×10^{11} Hz · 2.0×10^{57} **(Eq. 4.12)**

implies how many Planck volumes (galaxies) there were at the start.

This starting frequency is 3.35×10^{14} higher than f_{max} (f_{max} = 9.55×10^{53} Hz). Since the Planck's highest frequency is f_{max} and has a uniform maximum density of 2448, this higher background frequency must come from the resonance of a different structure. Because resonant frequency and mass follow one another, it seems logical to me that this larger structure is 3.35×10^{14} more massive than the uniform Planck volume. Using this idea I conclude that the number of Plancks in the total universe is 3.35×10^{14}. As stated earlier, an estimate of the number of galaxies in the visible universe is 6.1×10^{11}. This would mean that the volume of the total universe is approximately 549 times that of the visible universe. The total universe would then have a radius that is ~ 8 times bigger than the visible universe (113 billion light-years). Admittedly, not very useful as estimates go, but the theory is so simple and straightforward that it is fun to try.

Unification

As crazy as it may seem this new theory of matter, that stable

Matter With Electromagnetic Resonance

particles are self-sustaining space density bubbles, provides a new insight into the four fundamental forces of nature. Gravity, the electromagnetic force, the weak nuclear force and the strong nuclear force can be shown to be different aspects of space density using this theory. The theory predicts the force's relative strengths. This will be shown on pages 79-84.

Figure 4.3 shows a linear approximation for the density function throughout the proton as defined by this theory. As described in earlier chapters the proton is a two-tiered spherically resonant cavity with a peak relative density of 2.0 compared to free space. This maximum occurs ½ way into the spinning resonance. Because spin resonant particles have a radius that is just shy of ½ wavelength ($\lambda/2$), there is some excess density (non-spinning) that is outside of the particle. [*If all of the excess density was completely contained in a non-spinning resonant cavity, the particle radius would have been $\lambda/2$ away from the center.*] Because there is no physical or virtual short at the particle edge, the excess density present at the edge falls off as $1/r$ out to infinity. It is this relative density, peak inside vs. particle edge outside, that explains the four forces of nature and their relative strengths.

Gravity

Strangely enough, the one force that has defied unification in the past was the most straightforward for this book. The equation for the gravitational force is shown in Equation 4.13.

Matter With Electromagnetic Resonance

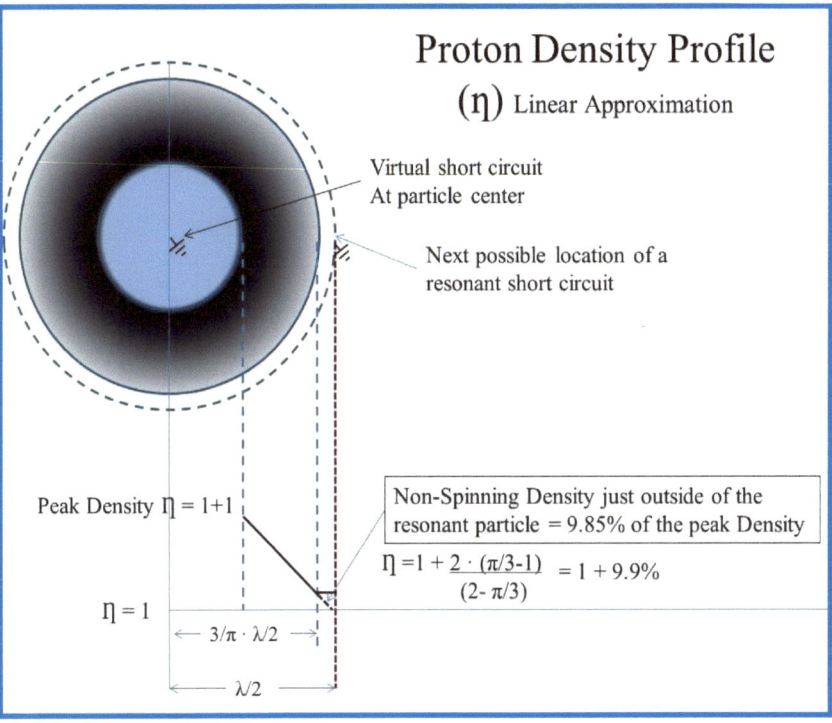

Figure 4.3 *Density profile of a proton (linear approximation). The relative excess density at the edge of the resonant particle is a little less than 10% of the peak excess density contained in the particle. This fractional density, outside to inside, explains the relative strengths of the four forces of nature.*

$F_{gravity} = G\,(m_1 \cdot m_2) \div r^2$ (Eq. 4.13)

Where $G = 6.67 \times 10^{-11}$ (Newton $\cdot$ meter^2 $\div$ kilogram^2)
m_1 and m_2 are two masses (kilograms) spaced r (meters) apart.
The equation for the electrostatic force is shown in Eq. 4.14.

$F_{electrostatic} = K_e\,(q_1 \cdot q_2) \div r^2$ (Eq. 4.14)

Where $K_e = 9 \times 10^9$ (Newton $\cdot$ meter^2 $\div$ coulomb^2)
q_1 and q_2 are two charges (coulombs) spaced r (meters) apart.

Matter With Electromagnetic Resonance

For two identical charged particles the ratio of the two forces can be written as shown in Equation 4.15.

$$F_{gravity} / F_{electrostatic} = (G/K_e) \cdot m^2 \div q^2 \quad \textbf{(Eq. 4.15)}$$

For two particles that are identical, assign the square root of this ratio to each.

$$(F_{gravity} / F_{electrostatic})^{1/2} = (G/K_e)^{1/2} \cdot m \div q \quad \textbf{(Eq. 4.16a)}$$

Using the energy equations, solve for charge q.
$$e = m \cdot C^2 = V \cdot Q \quad \textbf{(Eq. 1.9)}$$
$$q = Q = m \cdot C^2 \div V \quad \textbf{(Eq. 1.9a)}$$

Re-writing Eq. 4.16a, substituting for q.
$$(F_{gravity} / F_{electrostatic})^{1/2} = [(G/K_e)^{1/2} \cdot m \div m] \cdot V \div C^2 \quad \textbf{(Eq. 4.16b)}$$

The bracketed term looks like a constant fractional mass with a value of $(G/K_e)^{1/2} = 8.609 \times 10^{-11}$ **(Eq. 4.17)**

Referring to Figure 4.3, since there is fractional space density outside of the resonant particle and the relationship between space density and mass (mass $_{relative}$ = $\eta_{relative}^{10}$) is known. I can calculate what the outside space density would need to be to have a fractional mass of 8.609×10^{-11}. The tenth root of the fractional mass $[(8.609 \times 10^{-11})^{.1}]$ is = 0.0985, or **9.85% density**. This appears to be what the force of gravity is. It is the excess space density and mass contained outside of a spin resonant particle. This excess space density has the ability to slow down the speed of light [**Gravity.**].

Matter With Electromagnetic Resonance

The Strong Nuclear Force

The peak relative excess space density on the inside of the proton compared to the outside surface also explains the relative strengths of the Electrostatic and Strong Nuclear forces. Thinking about the density of space changing the speed of light, and the fact that this density is what provides the "forces", then looking again at figure 4.3 it can be shown that the excess charge density at radius = ½ way into the proton is [1/0.0985 = 10.15] times as dense as the charge density at the edge. The strong nuclear force must be 412.3 times stronger than the electrostatic force even though they both use the same electrostatic force equation.

This value is right in line with currently published relative strength values.

$$F_{electrostatic} \text{ (electrostatic)} = K_e (q_1 \cdot q_2) \div r^2 \quad \textbf{(Eq. 4.14)}$$
$$F_{electrostatic} \text{ (strong)} = K_e (10.15 \times q_1 \cdot 10.15 \times q_2) \div (r/2)^2 \quad \textbf{(Eq. 4.14a)}$$
$$F_{electrostatic} \text{ (strong)} = \underline{\textbf{412.3}} \cdot F_{electrostatic} \text{ (electrostatic)} \quad \textbf{(Eq. 4.14a)}$$

The Weak Nuclear Force
(controlling nuclear decay)

The weak nuclear force is also an electrostatic force. To see what is happening refer to Figure 4.4. Protons and neutrons are both two-tiered resonant particles that differ in their charge distributions in

Matter With Electromagnetic Resonance

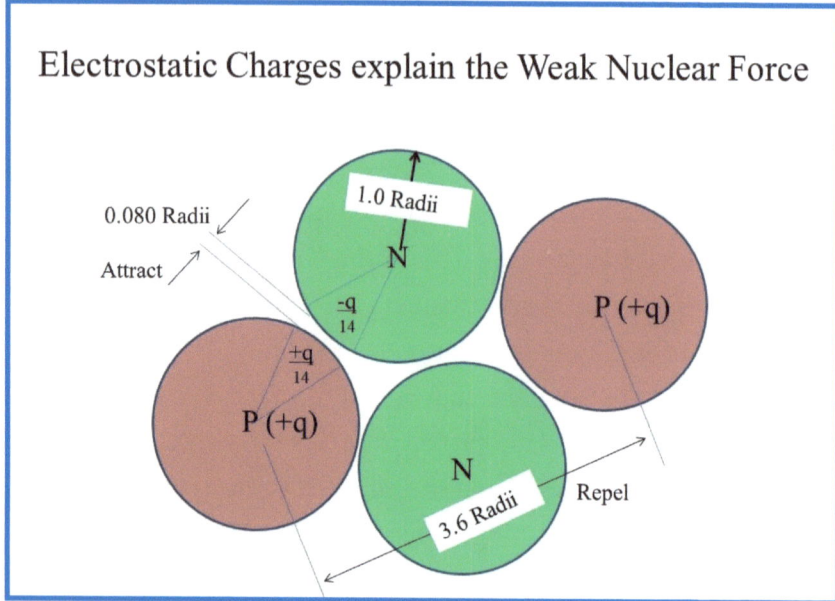

Figure 4.4 *Theory explains why the weak Nuclear Force has the strength that it has and how it holds the nucleus together.*

their outer tiers. The seven resonant cells on the outside of a proton are all positive. However, of the seven volumes (or fourteen half volumes) in the outer tier of the neutron, three of them are negatively charged and four are positive (See Figure 4.5). As stated in chapter I, this is a mathematically allowable arrangement that leaves the neutron with a net charge of 0.0 when viewed from a distance. When protons and neutrons are side by side however, the attraction of adjacent positive and negative cells can exactly cancel the repulsive force of a nearby proton. I won't show the math here but in terms of particle radii, when the opposite charges of (q ÷ 14) are 0.080 particle radii apart edge to edge, they have enough attraction to exactly cancel the repulsive force between two protons that are centered 3.6 particle

Matter With Electromagnetic Resonance

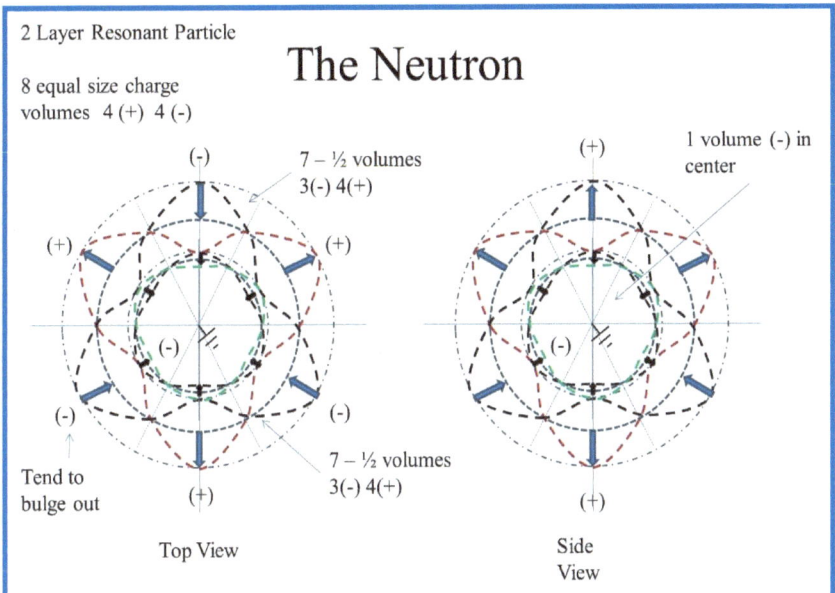

Figure 4.5 *Arrangement of positive and negative charges around the neutron provides a mechanism for protons and neutrons to stick together.*

radii apart (total proton charge q always looks as if it is coming from the center of the particle). This edge to edge distance occurs when the proton-neutron centers are almost exactly 1.0 resonant wavelength apart. If the particles are closer than (0.080 particle radii apart edge to edge) then the attractive force dominates.

Summary and Conclusion

This new theory successfully explains and describes the four forces of nature. They are all different aspects of the same thing, self-sustaining electromagnetic resonance and the density of space. Table 4.1 shows a comparison of the measured four forces

Matter With Electromagnetic Resonance

Comparison of Published vs. Calculated Strengths using this new Theory

	Windows2 universe.org	Georgia State University	SciencePark.etacude.com	New World Encyclopedia	New Theory Calculated Strengths
Strong	1	1	1	1	1
Electromagnetic	0.007	0.007	0.010	0.010	0.002
Weak	1.00E-09	1.00E-06	1.00E-07	1.00E-13	6.11E-09
Gravity	1.00E-38	1.00E-39	1.00E-38	1.00E-38	1.93E-39

Table 4.1 *This is a comparison of the measured relative strengths of the four fundamental forces at the level of the proton to the calculated strengths of the same forces using this new theory of matter.*

and their perceived strengths at the level of protons vs. the calculated strengths using the simple mathematical tools provided by this new theory. In my opinion this is an astonishing conclusion that adds more evidence that this theory is correct.

Based on everything that this new theory predicts it is clear to me that our present mental model of reality is incorrect. Space is not a large volume of nothing that is being stretched by the masses of matter that are floating in it! Rather, space is an actual substance (call it the Aether) with matter simply being denser parts of that same Aether. These denser parts are leftover excess density bubbles that precipitated outward during the Big Bang. They are self-sustaining due to the electromagnetic waves trapped and spinning inside of each stable particle.

Matter With Electromagnetic Resonance

Chapter Five
Black Holes

How to Build a Black Hole

Up to this point in the book, all of the particles that are stable, except the neutrino, are trapped Transverse Electromagnetic waves (TEM waves). These TEM waves are self-contained in spherically shaped resonant cavities. Transverse waves travel in a direction that is perpendicular to the direction of applied pressure (voltage). An example would be squeezing a balloon and having it bulge out to the sides, at right angles (90 degrees) to the direction of pressure. As stated earlier, these TEM particles are excess density bubbles that precipitated out of each expanding galaxy during the Big Bang. Each galaxy began as a uniform density Planck volume. Without too much variation, the amount of space contained in each stable bubble is approximately equal to 1.0 Q (0.7×10^{-45} *meter^3 for the electron*). It seems to me a logical

Matter With Electromagnetic Resonance

question might be; during expansion, what happened to any compressive waves? Because space is a material (I would call it the Aether) it should be able to support compressive waves. Where are they? A compressive wave travels in the direction of pressure. Sound is a compressive wave. Air molecules push against one another as the vibration expands outward from the source. Based on this new theory, any compressive waves contained in the early galaxy formation got trapped in a massive bubble surrounding the center of the galaxy. The bubble size and mass can vary depending on how much energy is contained inside, but the bubble wall thickness is the same as that of a neutrino (~ 1.6×10^{-35} meters), the Planck length. In my opinion, these structures are the Super-massive Black Holes that appear to be at the center of every galaxy. This is a natural outcome from this new theory of matter. I will describe the size, mass and other properties of these compressive wave structures.

The description that follows applies to only one kind of black hole. These are thin walled spherical shells with a virtual short circuit in the center. This theory also predicts another kind of black hole. For instance, if some of the massive TEM wave dark matter particles manages to gather together due to their collective gravity, they are small enough and massive enough to form an event horizon. An event horizon is a calculated non-physical radius surrounding a black hole. At the event horizon the escape velocity is the speed of light. Inside of that radius nothing, not even light, can escape the black hole's gravity. That kind of high density black hole will not be discussed here.

Matter With Electromagnetic Resonance

Low Density Super Massive Black Holes

If a non-spinning space density bubble with the volume of 1.0 Q tries to form at a frequency close to the electron frequency f_0 (1.236 x 10^{20} Hz), that particle will not be able to sustain its size using spin resonance. That bubble will expand until its wall thickness hits the Planck length of 1.6 x 10^{-35} meters. These compressive wave particles are what I believe physicists call the neutrino. The conservation of energy would say that this 1.0 Q volume will have the same total energy as that of an electron. By default, according to equation 5.1b it must also have the same internal voltage as the electron ~ 512000 volts.

$$\text{Energy}_{electron} = V_{electron} \cdot Q_{electron} = V_{neutrino} \cdot Q_{electron} \quad \textbf{(Eq. 5.1a)}$$
$$V_{electron} = V_{neutrino} \quad \textbf{(Eq. 5.1b)}$$

Because each 1.0 Q volume has the energy of an electron, the number of these potential volumes must be the same as the number of potential electrons. However, these neutrinos have a mass that is ~1.6 x 10^6 less than that of an electron. This means that unlike the spin-resonant particles which divide up the available mass in the expanding galaxy evenly, the compressive mass is ~1.6 x 10^6 less than the other particles. As I will show, compressive particles more than make up for having less total mass. Their gravitational strength is significantly higher than that of spin-resonant matter. The extra voltage pressure (512000 volts

Matter With Electromagnetic Resonance

vs. 0.32 volts mass-energy) in the neutrino does not show up as mass. Instead, the extra voltage pressure pushes ½ of the mass (or ½ of the density) outside of the particle (see figure 5.1a). Compared to visible matter, since ½ of the mass is outside of the particle's resonant radius, the density at the edge of the compressive particle is 9.47 times higher than 9.85% (% density

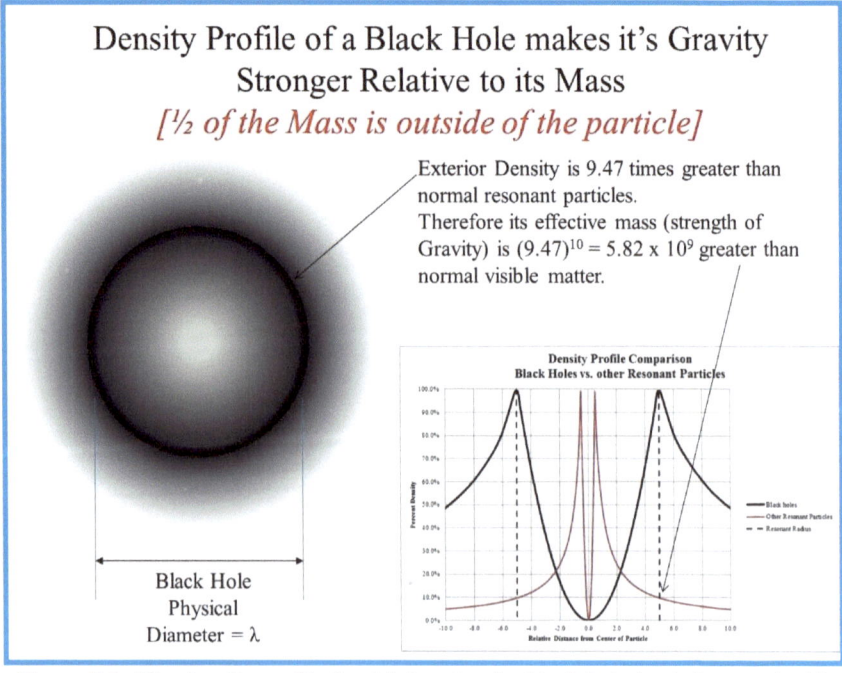

Figure 5.1a *The density profile for this low density black hole (and the neutrino) is different than that of normal visible matter. When the expanding pressure wave hits the Planck thickness, it bounces off and goes back toward the center forming a standing wave at the edge. This cancels charge and makes the density uniform across the wall thickness. Uniformity and extra pressure inside causes the density at the edge of the particle to be 9.473 times greater than that of a spin-resonant visible matter particle. Because of this the effective mass (gravity) is 5.82 <u>Billion</u> times greater than normal visible matter ($9.473^{10} = 5.82 \times 10^9$).*

Matter With Electromagnetic Resonance

at the edge of visible matter). Using the relationship between density and mass, this increases the neutrino's gravitational force (effective mass) by 5.82 x 10^9. If I use ½ of the density outside instead of ½ of the mass, I calculate a similar gravitational increase.

Gravitational Increase Using ½ Density Outside

For the voltage of the neutrino to be the same as the voltage of the electron, something has to happen to the impedance (density of space) that the neutrino sees (figure 5.1b). Looking at equations 5.2a and 5.2b, with Q being the same but the resonant frequency for the neutrino dropping, according to Ohm's law the impedance must go up. The impedance (density) goes up by the factor shown in eq. 5.3a. The extra "$\pi/3$" in the factor comes about because the compressive energy wave is not spinning but simply moving out radially. The compressive resonant radius is $\lambda/2$ instead of $[3/\pi \cdot \lambda/2]$. When the expanding wave slams into the Planck wall of minimum thickness, it bounces off. This standing wave cancels charge. Moving charge (current "I") goes to zero and the voltage doubles. That voltage factor of two shows up in the parentheses in equation 5.3b. To calculate the correct space density at the wall, equation 5.3b also contains an extra factor of two on the left hand side. The peak density drops by a factor of two when going from tapered to uniform (see figure 5.2). Because of the extra voltage pressure inside, the density at the wall for these compressive wave particles goes up not unlike

Matter With Electromagnetic Resonance

the density going up for the much smaller TEM resonant particles. The difference here is that, as stated before, the particle mass does not go up but rather its gravitational strength does. Its effective mass on the outside goes up.

$V_{electron} = Q_0 \cdot f_0 \cdot \Omega_0$ **(Eq. 5.2a)**
$V_{neutrino} = Q_0 \cdot (f_0 \div (f_0/f_n \cdot \pi/3)) \cdot (\Omega_0 \cdot (f_0/f_n \cdot \pi/3))$ **(Eq. 5.2b)**

Equation 5.3c calculates the increased peak space density at the Planck thickness wall. The density of space at the wall goes up to ~ 2.25. In terms of relative density, this is a little greater than the density of the proton.

$f_0/f_n \cdot \pi/3 = (1.236 \times 10^{20} / 7.7 \times 10^{13} \cdot \pi/3) = 1.681 \times 10^6$ **(Eq. 5.3a)**
$\eta_{neutrino} \cdot 2 = (1.681 \times 10^6 \cdot 2)^1 = 4.49$ **(Eq. 5.3b)**
$\eta_{neutrino} = (1.681 \times 10^6 \cdot 2)^1 / 2 = 4.49/2 = 2.25$ **(Eq. 5.3c)**

With a peak density of 2.25 at the neutrino edge and all of this showing up as increased gravity, it would be equivalent to a particle that has ~ 5.52 x 10^9 more gravitational mass than a neutrino. The earlier calculation using ½ mass showed a gravitational mass increase of ~ 5.82 x 10^9. These effective masses calculated using different approaches are both extremely high and similar enough that I believe there is a huge gravitational force increase for compressive wave structures. This large gravitational increase (~5.82 x 10^9) more than makes up for having less available compressive mass during the initial expansion of the galaxy.

Matter With Electromagnetic Resonance

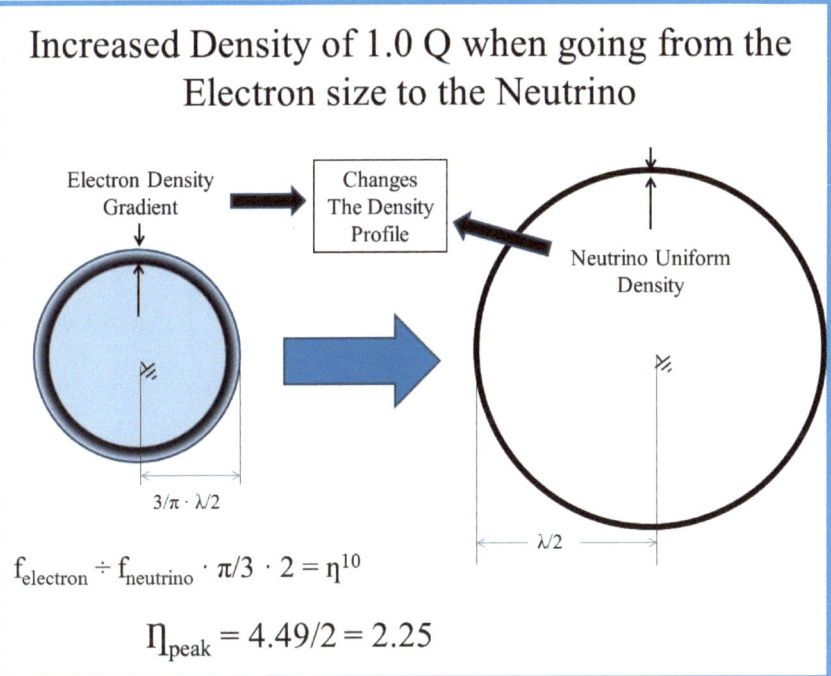

Figure 5.1b *The impedance (density of space) increases when a 1.0 Q volume blows up to the size of a neutrino. The density is uniform across the particle resonant edge. The neutrino peak density of (2.25) would make its gravitational strength ~ 5.52 x 10^9 greater than a particle of its mass would normally have.*

Actual Size (Physical Radius) of Super Massive Black Holes and Their Event Horizons

The equation for relative mass using the mass of the electron is restated below in Equation 1.2.

$m_1 = Q_1^2 \cdot \eta_1^{10} \cdot m_0$ **(Eq. 1.2)**
Where m_0 = mass of the electron
η_1 = new space density
Q_1 = amount of compressed space (for this neutrino case $Q_1 = Q_0$)
m_1 = new mass

Matter With Electromagnetic Resonance

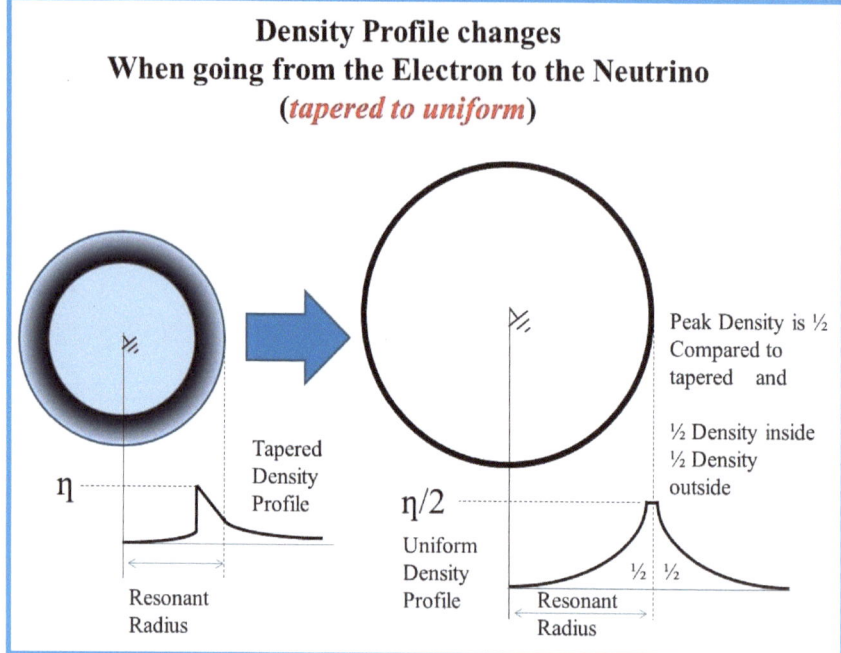

Figure 5.2 *Even though the density of space increases when a 1.0 Q volume blows up to the size of a neutrino, the changing density profile reduces the expected density peak by a factor of 2. Because the density is uniform at the particle edge and not tapered like the electron, ½ of the density is inside and ½ falls outside*

Because super-massive black holes are giant versions of neutrinos, I count neutrinos to calculate black hole gravitational masses. I will renormalize equation 1.2 into the forms 5.4a and 5.4b to calculate the effective (gravitational) mass of the neutrino and these related black hole structures. This will allow me to figure out actual super-massive black hole sizes and other properties.

Matter With Electromagnetic Resonance

mass $_{\text{gravitational, neutrino}}$ = $(N \cdot Q_0)^2 \cdot 5.82 \times 10^9 \cdot$ mass $_{\text{neutrino}}$ **(Eq. 5.4a)**
Where $N = Q_0 = 1$
mass $_{\text{gravitational, super-massive black holes}}$ = $N^2 \cdot 5.82 \times 10^9 \cdot$ mass $_{\text{neutrino}}$ **(Eq. 5.4b)**
Where N = number of Q_0 bubbles.

Therefore, super-massive black hole gravitational mass is proportional to N^2.

The conservation of energy must hold for the sum of N neutrinos.

$E = Q^2 \cdot f \cdot \Omega$ **(Eq. 5.5a)**
$E_{\text{neutrino}} = N^2 \cdot f_{\text{neutrino}} \cdot \Omega_{\text{neutrino}}$ (N = 1) **(Eq. 5.5a)**
$E_{\text{black hole}} = N \cdot (N^2 \cdot f_{\text{neutrino}} \cdot \Omega_{\text{neutrino}})$ **(Eq. 5.5b)**

Unfortunately, equation 5.5b is not valid as written. Since mass is proportional to N^2 (not N^3) and mass is proportional to energy, for 5.4b and 5.5b to be true, when N is greater than 1.0, the resonant frequency and the impedance of the black hole must each go down by the square root of N. The surface area of the

$E_{\text{black hole}} = N \cdot N^2 \cdot [f_{\text{neutrino}} / (N^{1/2})] \cdot [\Omega_{\text{neutrino}} / (N^{1/2})]$ **(Eq. 5.5c)**

black hole grows. The corrected equation is shown in 5.5c.

As the number of bubbles (N) grows, the black hole radius must increase to conserve energy.

We now have enough information to compute the radii and

Black hole mass $_{\text{gravitational}}$ = $N^2 \cdot 5.82 \times 10^9 \cdot$ mass $_{\text{neutrino}}$ **(Eq. 5.4b)**
Black hole radius = $N^{1/2} \cdot$ radius $_{\text{neutrino}}$ **(Eq. 5.6)**

Matter With Electromagnetic Resonance

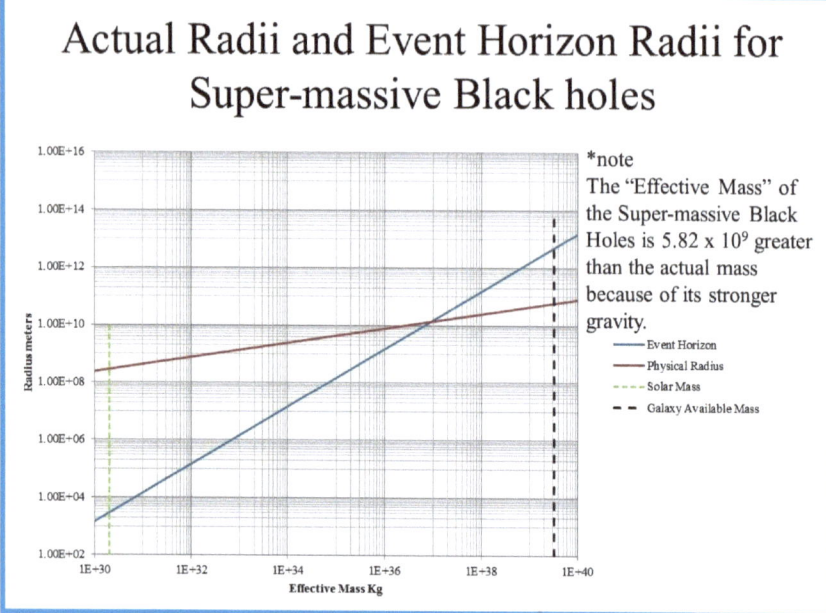

Figure 5.3 *A comparison of the actual physical radius of a super-massive black bole to its event horizon. These objects have 5.82×10^9 greater gravitational attraction than visible matter. This feature makes them look much more massive than they actually are.*

effective masses of these black holes.

Figure 5.3 contains a graph depicting the effective masses and sizes (radii) of these super massive black holes along with a calculation of their event horizons.

Conclusion

There are, at present, two different kinds of black holes predicted by this theory. One type is a physically small yet massive collection of heavy dark matter particles. These particles

Matter With Electromagnetic Resonance

are trapped, resonant, transverse electromagnetic waves. (That particular type seemed intuitive to me and was not described here other than that short description.) The other kind, which is described here, is a physically large trapped compressive wave structure. This compressive wave structure can be considered to have the same characteristics as the neutrino. This style of black hole has a wall thickness that has the Planck length ~ 1.6×10^{-35} meters. The effective (gravitational) mass is 5.82×10^9 times the gravitational mass of visible matter. As a consequence of the Big Bang I would expect some of the expanding wave energy to comprise of compressive waves. It seems reasonable to me that a massive compressive wave bubble is at the center of each expanding galaxy. This appears to be the case as astronomers look closely at all of the galaxies in the universe.

Chapter Six
The Aether

A Plausible Aether

After spending many hours thinking about stable particles of matter (see Figure 6.1) consisting of excess density bubbles held together by their own resonant waves in a material under compression (call it the Aether), it is inevitable that the mind will wander and start to wonder what that material might look like.

How might I construct a real substance that could behave in such a way? The material must be springy and compressible across a wide range of densities, yet seems to have limits on either end. It appears to have a maximum compressibility (the early universe $\eta=2448$) and a maximum stretch (the speed of light being constant today $\eta=1.0$). Both of these dimensions for stretch and compressibility seem to be related. They are somehow

Matter With Electromagnetic Resonance

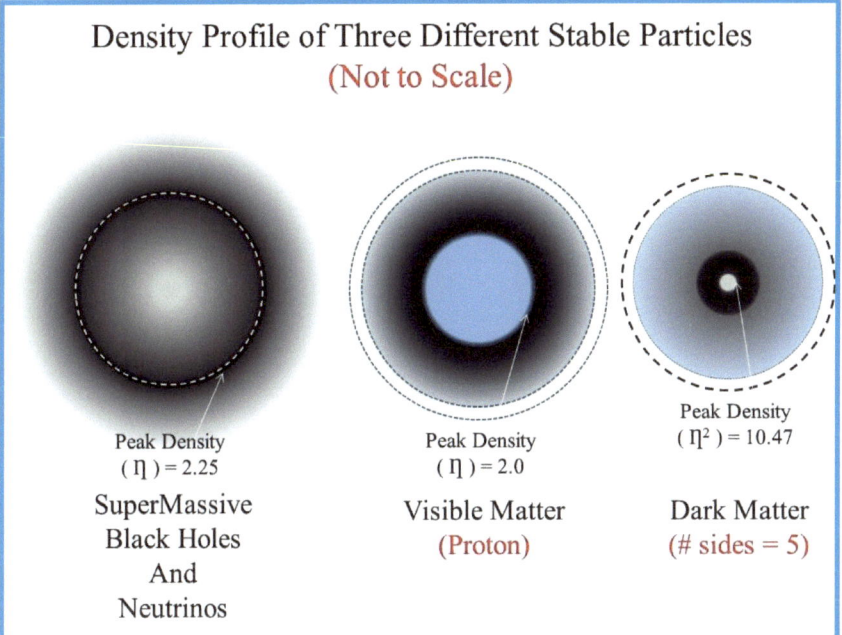

Figure 6.1 *Density profiles for three different stable particles. Supermassive Black Holes and Neutrinos are compressive wave particles. Visible and Dark Matter are spinning TEM wave particles. The dashed lines indicate the approximate location of ½ wavelength at the resonant frequency from the stable particle center if the index of refraction (density) were equal to 1.0.*

connected to the Planck Length of approximately 10^{-35} meters.

Figure 6.2 shows a picture of a possible unit cell comprising three orthogonal springy loops that could meet these requirements. Unimaginably thin wires arranged this way could sustain vibration (waves) and be able to fold in on themselves when compressed. Because the wires have size, at the maximum point of compression, the unit cell will be completely filled and hit a maximum density ($\eta \sim 2448$).

At the other end of the compression spectrum, when fully

Matter With Electromagnetic Resonance

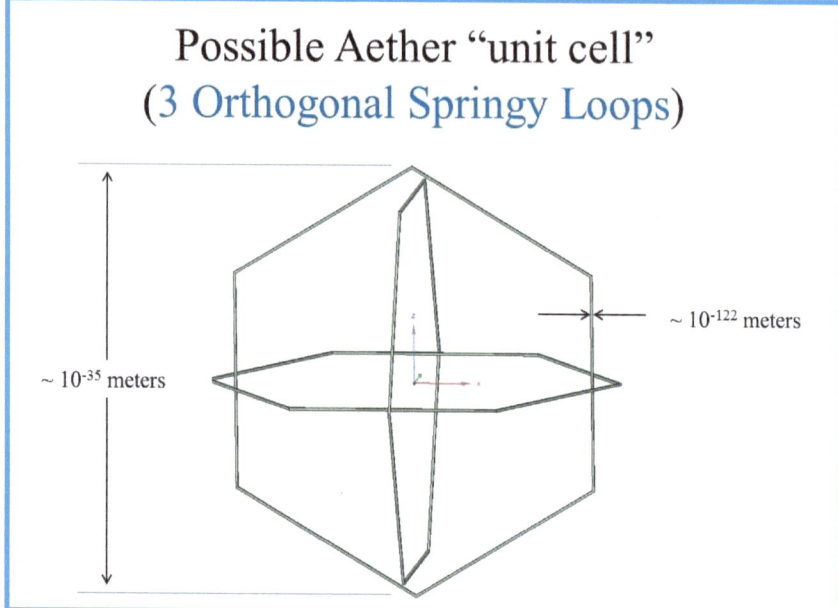

Figure 6.2 *If the Aether is a real material with size and shape and other properties, one possible configuration could be that as depicted above. A unit cell might consist of three orthogonal interlocking springy loops of a substance that can fold in on itself under pressure. With the approximate dimensions shown, an entire galaxy, when compressed, could fit inside of a volume that has edge dimensions on the order of the Planck length. Although this volume is extremely small it is not zero and therefore not a singularity.*

expanded, these wires will have reached maximum stretch. At that point they can stretch no further and the speed of light will have maxxed out and be constant. In the volume of the unit cell, everything that is not a springy loop is literally nothing. Since the cross-section of the wires in the loop is on the order of 10^{-122} meters, there is mostly nothing in the volume of the unit cell. This is the wire cross-section required to fit an entire galaxy of the Aether into this sized unit cell. Configuring the Aether this way is

Matter With Electromagnetic Resonance

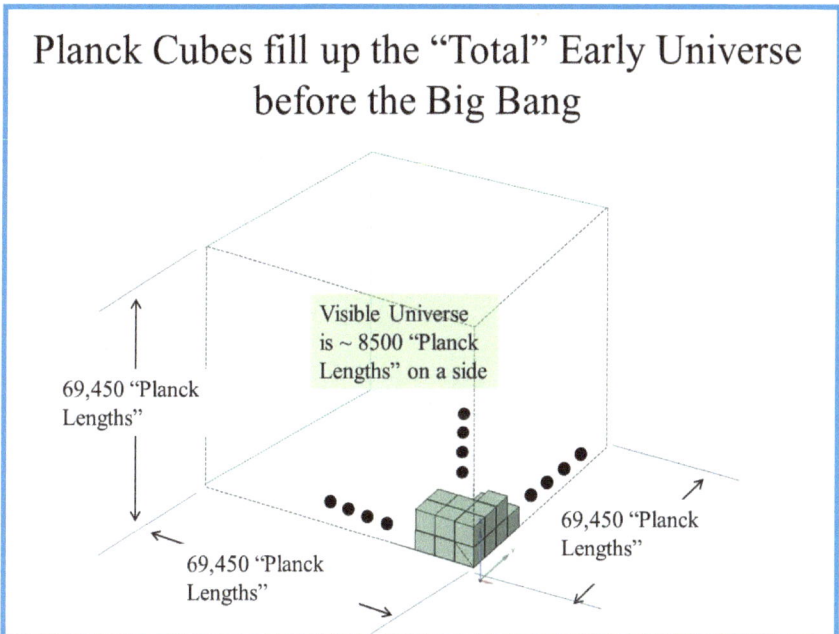

Figure 6.3 *Using the microwave background frequency of approximately 160×10^9 Hz and a calculated expansion of the universe of ~2×10^{57} one can calculate that there were ~3.35×10^{14} galaxies in the total universe. Before the Big Bang these Planck sized uniform density galaxies were sitting side by side in a tiny cube as shown above.*

not completely fanciful. It is fractal in its shape and orientation. Three dimensional spherical shapes abound in the universe both at the particle and the galactic level.

If the Aether is configured this way, tiny interlocking loops that can fold in on themselves under pressure, then the universe before the Big Bang might look like the picture in figure 6.3: Each galaxy compressed down to a cube that has edge dimensions the size of the Planck length. Based on the expansion of each Planck size to a galaxy size, calculated in an earlier chapter, and using the

Matter With Electromagnetic Resonance

measured microwave background frequency left over from the Big Bang, it is possible to estimate the <u>total</u> number of galaxies in the universe. That calculation says that the <u>total</u> universe has a volume that is ~ 545 times that of the visible universe.

This new theory that describes stable matter particles as self-sustaining resonant electromagnetic cavity waves did not come about using any existing theory as a starting point. Neither was it written to debunk any existing theory. Rather, it was put together to describe reality in a real, commonsense way based on my belief that the world around us is not an illusion as some recent theories have postulated. When I look at this theory in its entirety, I can see that aspects and pieces contained in this theory are touched on by some of the other theories being pursued today. It reminds me of the ancient Indian proverb that talks about several blind men touching an elephant. Each man holds on to a different part of the elephant and describes what he sees. One man holds the trunk, another the tail, another the ear, while yet another holds the leg. Each man describes a completely different animal. Clearly they are all in contact with the same animal, but from each point of view they see something different. I believe that this new theory is the elephant in the room. I will attempt to describe how some of the other theories appear to touch on aspects of this new idea.

Quantum Mechanics and Duality

Wave-Particle Duality

Matter With Electromagnetic Resonance

"The exhibition of both wavelike and particlelike properties by a single entity. For example, electrons undergo diffraction and can interfere with each other as waves, but they also act as pointlike masses and electric charges. The theory of quantum mechanics is an attempt to explain these apparently contradictory properties exhibited by matter.[11]*"*

Stable Particles are indeed waves and particles at the same time. Each particle consists of a resonant cavity density wave whose size is energy dependent. However, the particle also has a portion of its density wave outside of the cavity. The part outside trails off as 1/r to the edge of the universe and is what makes up charge and gravity. Particles do not take a stand when measured. They are always waves all of the time but have fixed, standing wave parts that have been called particles.

String Theory

"Any of a number of theories in PARTICLE PHYSICS *that treat elementary particles (see* SUBATOMIC PARTICLE*) as infinitesimal one-dimensional 'stringlike' objects rather than dimensionless points in* SPACE-TIME. *Different vibrations of the strings correspond to different particles.*[12]*"*

In Chapter Six, I have described a plausible Aether. The Aether consists of a wire frame of inter-connecting springy loops. The overall dimensions of the loops at the unit cell level are approximately the Planck length. The cross-section of the wires that make up the loops is extremely small such that a galaxy,

Matter With Electromagnetic Resonance

when fully compressed, can fit inside of a Planck sized unit cell. To me these loops sound a little like the loops of string theory.

10 Dimensional Space

"Superstring theory is a possible unified theory of all fundamental forces, but superstring theory requires a 10 dimensional spacetime, or else bad quantum states called ghosts with unphysical negative probabilities become part of the spectrum.[13]*"*

It is possible that this new theory's derived formula for the relationship between the density of space (index of refraction to the 10th power [η^{10}]) and resonant frequency and mass might have been touched on by the theories that claim that the universe has ten dimensions. In those other theories, three of these dimensions are the familiar ones, while the six others are wrapped up so tightly that they can't be observed. Time is also considered a dimension.

The Holographic Principle

"The holographic principle, simply put, is the idea that our three-dimensional reality is a projection of information stored on a distant, two-dimensional surface.[14]*"*

Reality is just a hologram and all particles of matter are just projections of information from the surface of a sphere. This idea seems to touch on the concept in this new theory that many of the

Matter With Electromagnetic Resonance

stable particles comprise incredibly thin spherical shells. Most of the mass is contained in the thin shells. Supermassive Black Holes and Neutrinos both have shells whose thicknesses are on the order of the Planck length 10^{-35} meters. Even the electron shell thickness is incredibly thin.

There is "nothing" out in space. (a commonly held belief for folks who don't like an Aether.)

These folks will be forgiven for thinking that there is nothing out in space because the fully expanded Aether unit cell ($\eta=1.0$) is about as close to nothing as you can get. These wires form loops that are ~10^{-35} meters in length, but have a cross-section of ~10^{-122} meters. The rest of the volume is nothing. And yet, this is the material that all of the electromagnetic waves are compressing as they propagate along throughout the universe.

Matter With Electromagnetic Resonance

m	(mass) The property of matter that gives it what is commonly called weight, a resistance to change in motion.
C	(speed of light 3×10^8 meters $\sec^{-1}$) The velocity of electromagnetic waves in free space.
h	h (Planck's constant 6.626×10^{-34} meter^2 kg $\sec^{-1}$) Describes the relationship between energy and frequency in particles.
V	Voltage (volts) The electromotive force.
Q or q	Charge (coulombs) A property of matter that causes it to experience forces in the presence of other charged particles.
I	Current (amps) Moving charge.
f	Frequency (Hz..... cycles/sec) Oscillation, rate of change.
λ	Wavelength (Lambda) The physical length of one complete oscillation at any given frequency.
Virtual ground	A "zero voltage" point or surface caused by the cancelling of equal and opposite voltages.
Ω	Impedance (Ohms) The ratio of the voltage to the current at any point in a system.
Density	The ratio of the mass divided by the volume.

Matter With Electromagnetic Resonance

η	Index of refraction (Space Density) A relative measure of the speed of light in any medium.
electron	Negatively charged stable subatomic particle.
positron	Positively charged subatomic particle.
proton	Positively charged stable subatomic particle.
neutron	Neutral quasi-stable subatomic particle.
Dark Matter	Unobserved hypothetical neutrally charged matter/particles.
Dark Energy	Hypothetical force expanding the universe.
neutrino	Extremely low mass particles that are hard to detect.
quarks	Unstable entities thought to be the building blocks of matter.
Resonant Cavity Modes	Stable physical shapes allowed by the laws of physics that are capable of storing energy in the form of oscillating waves.

Matter With Electromagnetic Resonance

Endnotes

[1] University of Michigan, Positron research website
NANOPOS "Positron physics"
[2] Cosmic Mismatch Hints at the Existence of a "Sterile" Neutrino, Scientific American, 20 February 2014, Clara Moskowitz
[3] theory.uwinnipeg.ca/mod_tech/node177.html or hyperphysics.phy-astr.gsu.edu/hbase/quantum/ debrog2.html
[4] The HAWC Gamma-Ray Observatory: Sensitivity to Steady and Transient Sources of Gamma Rays - HAWC Collaboration (Abeysekara, A.U. *et al.*) arXiv:1310.0071 [astro-ph.HE] HAWC Sensitivity to Diffuse Emission Petra Huntemeyer, Hugo Albert Ayala Solares, for the HAWC Collaboration
[5] The Vela-X region, (WMAP) data, GeV (Fermi) gamma-ray data, X-ray (ASCA) and TeV gamma-ray (H.E.S.S.) (from Fermi LAT Collaboration, 2010).Reference: H.E.S.S. observations of the Vela X nebula, H.E.S.S. Collaboration, F. Dubois et al., Proceedings of the 31st, ICRC, Lodz 2009

Matter With Electromagnetic Resonance

[6] Search for Gamma-ray Spectral Lines with the Fermi Large Area Telescope and Dark Matter Implications M. Ackermann et al. (23 Sep 2013) arXiv:1305.5597v3 [astro-ph.HE]

[7] Study of the Gamma-ray Spectrum from the Galactic Center in view of Multi-TeV Dark Matter Candidates - Belikov, Alexander V. *et al.* Phys.Rev. D86 (2012) 083516 arXiv:1207.2412 [astro-ph.HE]

[8] Wikipedia contributors. "Atomic radii of the elements (data page)." *Wikipedia, The Free Encyclopedia.* Wikipedia, The Free Encyclopedia, 29 Aug 2014. Web. 14 Sep 2014.

[9] The Large Distribution of. Galaxies Antonaldo Diaferio, page 1 (mariecurie.org/annals/volume1/diaferio.pdf)

[10] Wikipedia contributors. "Observable universe." *Wikipedia, The Free Encyclopedia.* Wikipedia, The Free Encyclopedia, 18 Nov 2014. Web. 19 Nov 2014.

[11] Dictionary.com, "wave-particle duality," in *The American Heritage® Science Dictionary.* Source location: Houghton Mifflin Company. http://dictionary.reference.com/browse/wave-particle duality. Available: http://dictionary.reference.com. Accessed: November 21, 2014.

[12] merriam-webster.com

[13] superstringtheory.com…Extra dimensions in string theory

[14] NOVA, Holograms, Black Holes, and the Nature of the Universe, By Kate Becker on Tue, 15 Nov 2011

www.ingramcontent.com/pod-product-compliance
Lightning Source LLC
Chambersburg PA
CBHW040221220526
45473CB00001B/66